BECOMING AN ENGINEER IN PUBLIC UNIVERSITIES

Palgrave Studies in Urban Education

Series Editors: Alan R. Sadovnik and Susan F. Semel

Reforming Boston Schools, 1930–2006: Overcoming Corruption and Racial Segregation
 By Joseph Marr Cronin (March 2008)

What Mothers Say about Special Education: From the 1960s to the Present
 By Jan W. Valle (March 2009)

Charter Schools: From Reform Imagery to Reform Reality
 By Jeanne M. Powers (June 2009)

Becoming an Engineer in Public Universities: Pathways for Women and Minorities
 Edited by Kathryn M. Borman, Will Tyson, and Rhoda H. Halperin (May 2010)

Multiracial Urban High School: Fearing Peers and Trusting Friends
 Susan Rakosi Rosenbloom (forthcoming)

The History of "Zero Tolerance" in American Public Schooling
 By Judith Kafka (forthcoming)

Becoming an Engineer in Public Universities

Pathways for Women and Minorities

Edited by
Kathryn M. Borman, Will Tyson, and
Rhoda H. Halperin

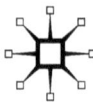

BECOMING AN ENGINEER IN PUBLIC UNIVERSITIES
Copyright © Kathryn M. Borman, Will Tyson, and Rhoda H. Halperin, 2010.
Softcover reprint of the hardcover 1st edition 2010 978-0-230-61935-7
All rights reserved.

Cover Image by Gyroscopic Studios

First published in 2010 by
PALGRAVE MACMILLAN®
in the United States—a division of St. Martin's Press LLC,
175 Fifth Avenue, New York, NY 10010.

Where this book is distributed in the UK, Europe and the rest of the world, this is by Palgrave Macmillan, a division of Macmillan Publishers Limited, registered in England, company number 785998, of Houndmills, Basingstoke, Hampshire RG21 6XS.

Palgrave Macmillan is the global academic imprint of the above companies and has companies and representatives throughout the world.

Palgrave® and Macmillan® are registered trademarks in the United States, the United Kingdom, Europe and other countries.

ISBN 978-1-349-38207-1   ISBN 978-0-230-10682-6 (eBook)
DOI 10.1057/9780230106826

Library of Congress Cataloging-in-Publication Data is available from the Library of Congress.

A catalogue record of the book is available from the British Library.

Design by Newgen Imaging Systems (P) Ltd., Chennai, India.

First edition: May 2010

10 9 8 7 6 5 4 3 2 1

As we applied the finishing touches to this volume, we lost one of our coeditors. Rhoda Halperin passed away April 10, 2009. Rhoda was not only a coeditor, she was a colleague and friend who will be deeply missed. Rhoda believed strongly in the importance of this research and supported our team, particularly through her mentoring of graduate students. We regret that she did not get to see this work in print.

Contents

List of Figures ix
Acknowledgments xi
Series Editors' Foreword xiii

One	Introduction: The Scarcity of Scientists and Engineers, a Hidden Crisis in the United States *Kathryn M. Borman, Rhoda H. Halperin, and Will Tyson*	1
Two	Producing STEM Graduates in Florida: Understanding the Florida Context *Bridget A. Cotner, Cassandra Workman Whaler, and Will Tyson*	21
Three	To Stay or to Switch? Why Students Leave Engineering Programs *Will Tyson, Chrystal A.S. Smith, and Arland Nguema Ndong*	53
Four	Pedagogy and Preparation: Learning to be an Engineer *Rebekah S. Heppner, Reginald S. Lee, and Hesborn O. Wao*	81
Five	Program Climate: Engineering Social and Academic Fit *Hesborn O. Wao and Reginald S. Lee*	105
Six	Program Culture: How Departmental Values Facilitate Program Efficacy *Susan Chanderbhan Forde, Cynthia A. Grace, and Bridget A. Cotner*	127
Seven	Making the Transition: The Two-to Four-Year Institution Transfer Experience *Cassandra Workman Whaler and Jason E. Miller*	147
Eight	Voices from the Field: Strategies for Enhancing Engineering Programs *Kathryn M. Borman, Will Tyson, and Cassandra Workman Whaler*	173

References 191
List of Contributors 199
Index 203

Figures

1.1	Multilevel Model of Student Retention	3
2.1	Florida Bachelor's Degree Recipients	22
2.2	Florida Engineering Degrees	23
2.3	Five Florida Public Universities	28
2.4	Florida A&M	30
2.5	Florida State University	30
2.6	Florida International University	37
2.7	University of Florida	42
2.8	University of South Florida	46
5.1	Means, Standard Deviations, and Correlation among Theorized Climatic Measures	113
5.2	Means, Standard Deviations, and Agreement on Climate Measures among Students by Department and Programs	114
5.3	Means, Standard Deviations, and Item Loadings on Data-Driven Climate Factors	117
7.1	Total Florida Enrollment, 2006–2007	148
7.2	2006 Florida Community College System Transfer Students by Race	149
7.3	2006 Florida State University System	150
7.4	Percent Pell Grant Recipients by Enrollment sand Ethnicity in Florida	151
7.5	Percent Transfer and Engineering Transfer, 2006	152

Acknowledgments

The NSF-sponsored study, "The Effects of College Degree Program Culture on Female and Minority Science, Technology, Engineering and Mathematics (STEM) Participation," would not have been possible without the support and guidance of colleagues and friends over the last several years. We would first like to thank the institutions who participated in this research: University of South Florida, University of Florida, Florida A&M, Florida State University, Florida International University, University of Central Florida, Florida Atlantic University, Florida Institute of Technology, Embry Riddle Aeronautical University, St. Petersburg College, Hillsborough Community College, Santa Fe Community College, Tallahassee Community College, Miami-Dade Community College, and Broward Community College. Their continuous support of our project enabled us to conduct our research with considerable ease, and we look forward to any opportunity to collaborate in the future. Our heartfelt gratitude goes to the over 2,300 research participants in our study. They not only provided the data allowing us to complete our study but also, more importantly, provided insight into the daily lives of engineering majors and shared impressions of faculty, administrators and staff in engineering departments. We would also like to thank the members of our Advisory Board who provided valuable insight during the formative stages of our research and throughout the research process. Thank you to Cheri Ostroff, Melanie Cooper, Roger Seals, Michael Gaines, Jeylan T. Mortimer, and Mark Lewine. Thank you, Cheri, for conducting the evaluation of our project.

In addition to the authors of this book, the efforts of many people contributed directly to this research. We would like to thank the following for their dedication to this project: Mary Ann Hanson, Heather Ureksoy, Eva Fernandez, Ted Micceri, Melissa Rivera, Caroline Peterson, Cynthia A. Grace, Sam Arcangeli, Elaine Mueninghoff, Julia Fraser, Jaime Davis, Tasha-Neisha Wilson, Diane Cotsirilos, Anna Tolentino, Jennifer Hunsecker, Ashley Nixon, Sandra Gonzales, and Ana Torres.

The authors acknowledge the generous support of the National Science Foundation through NSF Grant # 0525408: "The Effects of College Degree Program Culture on Female and Minority Science, Technology, Engineering and Mathematics (STEM) Participation." We also appreciate Susan Hixson's guidance and support throughout our work. Any opinions, findings, and conclusions or recommendations expressed in this volume are those of the authors and do not necessarily reflect the views of the National Science Foundation.

Finally, the authors would like to thank the Palgrave Macmillan team including, Samantha Hasey, Assistant Editor, and Alan Sodovnik, Series Editor.

Series Editors' Foreword

> *... Well, I listened to my mother and I joined a typing pool*
> *Listened to my lover and I put him through his school*
> *If I listen to the boss, I'm just a bloody fool*
> *And an underpaid engineer*
> *I been a sucker ever since I was a baby*
> *As a daughter, as a mother, as a lover, as a dear*
> *But I'll fight them as a woman, not a lady*
> *I'll fight them as an engineer!*
> —Peggy Seeger, "I'm gonna be an engineer," 1970, Stormking Music

Forty years after Peggy Seeger, the sister of folk singer Pete Seeger, lamented in lyric the obstacles facing women, both as prospective engineers and engineers, the paucity of women and minority engineers remains a critical problem. *Becoming an Engineer: Pathways for Women and Minorities* summarizes the results of a three-year National Science Foundation (NSF)–sponsored investigation on limits and possibilities of higher education programs in increasing access for women and minorities into the STEM professions in science, technology, engineering, and mathematics. Through an examination of the ways in which the culture and organizational conditions of higher education may or may not result in the successful completion of undergraduate engineering degrees for those traditionally underrepresented in engineering careers, the book provides essential sociological evidence for providing increased access, opportunity, and results. Borman and her colleagues examined programs at two- and four-year public institutions of higher education in the State of Florida, the results of which they argue may be applied nationally. These institutions vary by size and scope of available undergraduate programs and also in their relative success in graduating women and minority students and indeed all groups of students in science, technology, engineering, and mathematics (STEM) majors.

Using a mixed-methods design, including qualitative methods, such as observations, interviews, and focus groups and quantitative methods, primarily surveys, this book contributes to the knowledge base concerning education and STEM careers, but also more broadly to the literature on organizational culture and the ways in which it affects important higher education outcomes. It provides significant insights about school climate as it is reflected in university, college, and departmental policies and goals as well as in the administrative and academic supports provided to engineering undergraduate students. It examines factors that influence engineering undergraduate students' experiences as they seek to "fit" in and accomplish their academic goals, student preparation and curriculum issues that challenge both engineering undergraduate students and faculty responsible for preparing students for careers in civil and electrical engineering. The authors examine the community colleges as pathways to STEM programs at four-year colleges and universities and how student experiences affect their success as they transfer into four-year institutions. In addition, they explore the factors that result in persistence or attrition from these programs. Finally, the book provides policy implications for universities engaged in educating future engineers and research based best practices to increase the number of minority and women engineers.

This book illustrates the value of social science research in identifying the remaining obstacles to increased participation and the best practices for increasing the number of women and minority engineers. Finally, it illustrates the importance of a sociological perspective that emphasizes organizational processes, culture, and climate rather than a cultural deficit model that emphasizes group limitations, in understanding the limits and possibilities of colleges and universities in providing pathways for success.

ALAN R. SADOVNIK
SUSAN F. SEMEL

Chapter One

Introduction: The Scarcity of Scientists and Engineers, a Hidden Crisis in the United States

Kathryn M. Borman, Rhoda H. Halperin, and Will Tyson

> *A doctor impacts one life at a time, that one patient that they're seeing. But engineers will impact millions with the technology you come up with…*
> —University of Florida Black female engineering student quoting a University of Florida Black administrator

Introduction

Engineers are the unsung heroes of the twenty-first century. Engineers build the physical and technical infrastructures that laypeople often take for granted. Most of us do not think about civil engineers as we drive over bridges or about the work of electrical and computer engineers when we use our Blackberries. However, currently, U.S. high schools and universities do not produce enough students who pursue and persist in engineering careers. Our research is motivated by this crisis: a scarcity of new scientists in the United States. There is an urgent need for highly educated workers in science, technology, engineering, and mathematics fields (STEM). Employment in STEM occupations during the current decade is expected to increase three times faster than employment in all remaining occupations (National Science Board, 2002). In addition, 25% of U.S. scientists and engineers will reach retirement age by 2010 (Building Engineering Science Talent Report, 2004). Important new opportunities emerging at the intersection of information technology, life sciences, materials sciences, and engineering are critical to the recovery and continued success of the U.S. economy. In fact, the Bureau of Labor Statistics projects that our greatest needs in the future will be in computer-related fields that drive innovation. To understand what can be done to facilitate the success of engineering undergraduates we conducted interviews, surveys and observations at several large public universities to gain first-hand, in-depth knowledge about engineering education. Future civil and electrical/computer engineers are key voices in this study along with professors, staff members, and administrators who work with them.

Concurrent with high demand for increasing the engineering and STEM labor force is the changing composition of U.S. workers. Women and minorities are the fastest growing demographic groups among working people currently employed. By 2010, women will earn more degrees than men at every level of higher education from the associate degree to the doctorate. By 2015, the nation's undergraduate population is expected to expand by more than 2.6 million students, 2 million of whom will be students of color. To produce sufficient numbers of workers and remain competitive in the global economy, the United States must find ways to encourage women and minorities to enter engineering and the sciences more generally (Committee on Equal Opportunities in Science and Engineering Report, 2000, p. 41). The sciences are critically important to the United States as competition for highly skilled workers intensifies globally. The urgent need for engineers and scientists presents enormous challenges to the nation's educational systems. Statistics on science degree attainment among women illustrates this. Although by 2010 women will earn more degrees than men at every level of higher education, they continue to lag behind men in science and engineering degrees. For instance, women earn only 19% of physics degrees, 2% of computer science degrees, and 18% of engineering degrees (National Science Foundation, 2000). For minority students, the numbers are equally troubling. In 1996, African-Americans, Hispanics, and Native Americans made up 23% of the 18 to 29 year-old age group (US Department of Education, 2000), but these minority groups taken together accounted for only 17% of the associate's degree recipients and only 14% of bachelor's degree recipients in science and engineering in this same time period. As such, African-Americans,[1] Hispanics and Native Americans are considered under-represented minorities in science, technology, engineering, and mathematics.[2] To address the crisis, the National Science Foundation (NSF) created research support programs, supported by Congressional mandate, for STEM research at the postsecondary level.[3]

Theoretical Frameworks

The theoretical framework driving the research undertaken by our interdisciplinary team over the past several years reflects a multi-level, multi-factor model encompassing institutional structures, programs and individual actors. This model is displayed in figure 1.1. This figure shows the set of factors contributing to the major outcome examined in this study: the capacity of the four public state university engineering programs in our study to successfully graduate civil and electrical engineers.

The work of sociological theorists Anthony Giddens and Pierre Bourdieu figure prominently in this book and help frame adaptive strategies for the success of women and minorities in the historically white male dominated discipline of engineering. This book examines strategies employed by individual students to better navigate engineering as well as strategies employed over the years by faculty, administrators, and staff to encourage their students to persist and complete an engineering degree. This book also describes successful institutional factors and puts forward recommendations for institutional change. Here, the study of practice complements the study of structures with the Florida State University System itself a serving as a microcosm of engineering programs throughout the country.

The research questions and analytical methods for this project are guided by three basic conceptual frameworks used to understand the experiences of women and minorities as underrepresented groups in engineering. Together, these three frameworks constitute a multidisciplinary, globally oriented approach (Appadurai, 2001; Sassen, 1998). Political economy, practice theory, and person-environment fit all inform the understanding

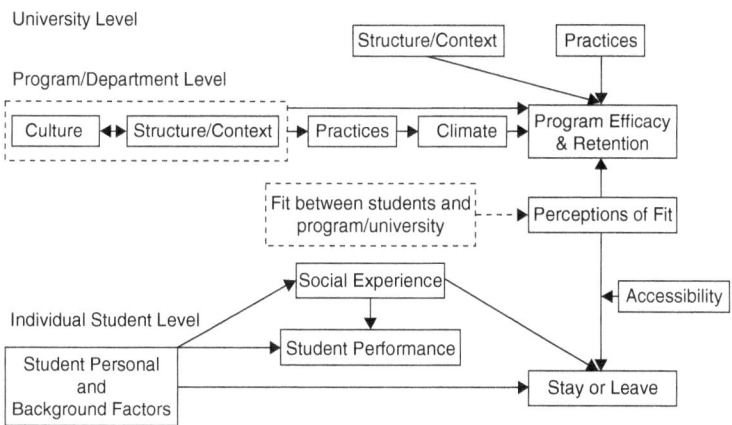

Figure 1.1 Multilevel Model of Student Retention
Source: This is an adaptation by Cheri Osteroff.

of culture and climate. These three conceptual frameworks help in understanding successes as well as obstacles in engineering experienced by women and minorities at the institutional, program/department, and individual levels.

Political Economy

The conceptual framework of political economy draws from the disciplines of anthropology and economics. Political economy focuses on material resources in terms of their production, distribution, consumption, and exchange across cultures and over time (e.g., Halperin, 1994). Ecology dovetails with political economy by examining relationships among humans, their physical and institutional (social and economic) environments, and their abilities to adapt, survive, and succeed in often unfamiliar and uncomfortable environments. Political economy focuses on the economic and political contexts (or structures) within which groups exist. At the time of this writing, the world economic crisis is uppermost in everyone's consciousness. The availability of financial aid, for example, disproportionately impacts students of lower socioeconomic status. Unless institutions of higher education, themselves facing budget cuts, develop creative solutions, women and minority students may be particularly hard hit.

In this project, we look at women and ethnic/racial minorities in the current historical moment. Changes in the political and economic environment will affect the abilities of women and minorities to succeed as engineers. The question of how students maintain a sense of cultural, ethnic and class authenticity and still succeed academically is quite complex. At the very least, mixtures of minority, working class and academic cultures produce forms of hybridity that have not been analyzed carefully (Garcia Canclini, 1995). Hybridity refers to the mixture of multiple characteristics in one person. An engineering student who is an African American woman can be active in the black community, smart and highly accomplished academically without losing her authenticity as a woman and without succumbing to the social isolation associated with a "nerdy" academic culture. She does not have to give up her original identities in order to become something else, in this case, an engineer.

Political economy focuses on social structure and its relationship to individuals (Giddens, 1979). Social structures obviously make a difference in the experience of members of society, for example women and minorities. These structures may constrain their ability to successfully transition into and persist in engineering programs. Political economy is particularly concerned with issues of history, social class, socioeconomic positioning, and shifts in positioning, and the resultant economic disadvantages. For example, historical race and class lines are still realized in the social geography of American cities. Despite advances in civil rights and social attitudes toward segregation, cities are still largely segregated by race and class resulting in unequal distribution of educational resources. Inner city and rural schools often lack experienced teachers, challenging coursework, and adequate textbooks required to achieve the quality of education existing in more affluent areas. The political economy of poor schools places students attending them at a disadvantage compared with their more affluent counterparts when entering higher education. The opportunity to take advanced mathematics and science courses necessary for sufficient academic preparation directly impacts persistence toward an engineering degree.

In addition, we saw throughout our research that some engineering buildings at the public universities where we conducted our research have spacious and well equipped study facilities and are well-resourced; others require students to struggle for a quiet space and computer access. These differences in opportunity and access create circumstances that may not be conducive to student success, in this case, the success of women and minorities in engineering. As a result, women and other underrepresented minorities can be understood as disadvantaged in comparison with white males because of their social and political economic positions rather than because of inherent intellectual deficiencies, as has been argued in the past.

Theories of political economy allow researchers to be aware of how society impacts experiences of women and minorities in engineering and the strategies engineering students use to adapt and modify structural limitations they may face in engineering programs. How can engineering programs modify their cultures and climates to hasten the process of producing more engineers, especially women and minorities? How can the structural characteristics of engineering programs in this set of Florida public universities, make traveling the pathway to becoming an engineer, not just easier, but more and more possible for more and more individuals? To understand how one's political economic position is lived and how people negotiate the structural constraints they may face, we introduce practice theory.

Practice Theory

Practice theory is drawn from the work of Pierre Bourdieu (1977), rooted in the disciplines of sociology and anthropology. This approach allows us to focus on individuals and their positions in different institutional contexts; for example, how much cultural and symbolic capital they bring with them into engineering. Many students forge important relationships that allow them to gain the social currency, or social capital, necessary to traverse their programs. Student perceptions of their own power and agency over their educational trajectories are critical to their success (Foucault, 1980; Gramsci, 1971). First-generation minority college students, especially those who grew up in working class families with little education enter college with much less cultural and financial capital than students with long college legacies in their families. Many first-generation college students enter college with little financial or academic help from family members.

To highlight the importance of capital, approximately 25% of female undergraduate engineering students compared to 11% of male students interviewed in this study reported that their fathers were engineers. As shown in anthropological and sociological studies, women are brought up believing they are not good at math. Our research finds that women having an engineer as a parent have the advantage of knowing how important math is to earning an engineering degree. Having a parent with the knowledge of what it takes to become an engineer is important for women to mitigate cultural and institutional barriers they face in becoming engineers. Without such social capital women may chose another major with fewer math requirements, bolstering their chances of graduation but at the same time excluding them from the sciences. One Hispanic female mentioned that her female friends, "they didn't really understand all the math, so a lot of them just went into business [as a major]." This finding is reminiscent of Paul Willis' seminal book *Learning to Labour* that uncovers how disadvantaged students make choices that may seem to be in their best interests but instead maintain the social class status quo.

Willis (1977) analyzes issues confronting working class boys in England being prepared by the secondary education system for work on shop floors, not college. In effect, manual laborers reproduce themselves, generation after generation, first by being discouraged from advancement to college in secondary schools, and secondly by accepting their positions as subordinates in the capitalist system, now in the twenty first century, a system of global capitalism. This book demonstrates adaptive strategies used by successful engineering students in the face of adversity. In addition, student support can come from within institutions themselves. The interaction between the higher education system, including administrators, faculty and staff, and the students in engineering programs can further be explained by exploring person-environment fit, a concept drawn from industrial-organizational (I/O) psychology.

Person-Environment Fit

Person-environment fit examines how individuals relate to key organizational structures. The notion of fit is linked to organizational culture because the more congruent an individual's values are with the prevailing organizational culture, the more likely that individual is to stay and feel satisfied with the organization, in this case the engineering department in which the student is enrolled. Within I/O psychology, the definition of culture involves shared assumptions and values that manifest themselves in policies and practices at the institutional level, be it engineering college or department. Important in this definition is the notion that culture is shared by members of an organization and affects how individuals feel they fit within that organization. Climate is defined as the interpretation and experience of culture at the individual level. Factors related to climate create environments conducive to student success in the sciences, again, keeping agents' differential power in mind. For example, white male students at the University of Florida, with long legacies of college attendance and graduation with engineering degrees, have relatively more power, than first generation in college community college students from minority backgrounds.

Culture and Climate

Uniting the three conceptual frameworks, political economy, practice theory, and person-environment fit, is the concept of culture as shared perceptions, values, and assumptions along with a focus on practices within different institutional contexts and program environments. Since the main characteristics of underrepresented groups are relative powerlessness and inexperience, here, with engineering programs in institutions of higher education, we are dealing with disempowered populations. Two additional concepts drawn from the social sciences are also critical to this study: culture and climate.

Culture certainly matters at all levels of the university and affects retention most dramatically. Whether or not students enter the program with clear understandings about what is expected of them, the difficulty of their courses and the need to participate in collaborative relationships with peers, graduate students, and faculty also impacts retention. Two year community colleges clearly offer a "small is beautiful" model of education with classes of 30 rather than 300 in size. A smaller scale in turn leads to easier access to help from faculty and, certainly, a different kind of classroom experience. Clearly student personal characteristics— academic skills, abilities, goals and interests, finances—play important roles in student success. In addition, "campus ecologies," physical learning spaces, technology, infrastructures, and food availability, as well as facets of student performance such as course taking, grades, and timely progression toward the degree, are other important factors.

Climate is the experience of culture by groups of individuals. This conceptualization foregrounds the importance of student perceptions of engineering programs they attend. How students traverse their undergraduate careers, given the supports they have or create as well as the constraints they face, is critical to understanding the production of engineering undergraduate students. Importantly, students from different backgrounds may interpret departmental culture and climate differently, and for women and underrepresented minorities departmental culture and climate may not be conducive to retention to graduation. Indeed, previous research has demonstrated a "chilly" classroom and campus climate for women (Seymour & Hewitt, 1997). While the notion of a chilly climate is not without debate, many researchers agree that women and minority students may have markedly different college experiences (Allan & Madden, 2006; Salter, 2003). In studying students who have been historically at a disadvantage due to race, class, or gender, it becomes critical to understand why women and underrepresented students are not graduating at the same rate as their white male counterparts. These conceptual frames provide windows on the positions of women and minorities along engineering pathways as they transition from the early years of college and, often large, introductory courses, to smaller, upper level classes needed for graduating as engineers. In short, we have sought to uncover which departmental cultural and climatic factors are most important in predicting student success in terms of student retention.

Program Efficacy

Many factors affect student retention. The model, shown previously in Figure 1, informs our central research objectives, to identify aspects of culture and climate of STEM programs that affect retention of engineering undergraduates, and to understand how these factors affect underrepresented groups including minority students and women. This research examines how well different public institutions in Florida succeed in providing opportunities in engineering to students often underrepresented in the STEM disciplines. Similar research uses simple retention or graduation statistics for these purposes. We concur with Astin (1993) who contends that retention rates can be misleading because more than half the variance in retention rates can be directly accounted for by differences in types of students enrolling. In other words, student quality is a key determinant of student retention. In fact, some institutions with "high" retention rates should have even higher rates, given the high achieving students these universities admit. For example, students in need of remediation, as is the case for some students admitted to engineering programs attending Florida institutions in our study, seek assistance largely by working with their classmates as opposed to taking remedial coursework. In calculating graduation rates, most experts allow a period of from four to six years following matriculation for students to graduate from their respective programs.

So-called "simple" graduation rate statistics are anything but simple when attempting to compare institutions that vary in their selectivity and criteria for admission (Horn, 2006). Almost 60% of undergraduates attend more than one institution during their student careers, including students who supplement their curriculum with community college courses, and students who "swirl" between four-year universities and two-year colleges. The increase of dual-enrollment, credit-by-examination, and summer enrollment in the 1990s make singular university graduation rates a less effective metric of institutional effectiveness (Adelman, 2004). Emerging trends including online courses and other distance learning options must encourage researchers to be more creative in understanding the diverse and often circuitous pathways to bachelor's degree attainment.

In light of using problematic *retention rates* to compare institutions, this chapter introduces the concept of *program efficacy* to better describe the progress of underrepresented students in a specific higher education environment. Program efficacy is a normative term referring to programs in which female, Black, and/or Hispanic students exhibit greater progress toward degree attainment than underrepresented students in other programs as well as the general student population across the institution in question. In programs with low efficacy, underrepresented students do not progress toward degree attainment at a comparable rate with peers outside the program. The method for calculating program efficacy is described in chapter two along with descriptions of program efficacy within each engineering program.

Research Team

The Alliance for Applied Research in Education and Anthropology (AAREA) is housed in the University of South Florida (USF) Anthropology Department and led by Dr. Kathryn Borman. The AAREA research team is both interdisciplinary and multicultural, comprised of a diverse group of faculty and graduate students: anthropologists, sociologists, psychologists, women and men, senior scholars and graduate students.

Through the years, AAREA researchers have pursued a research agenda focused on issues of equity, investigating access to challenging and rigorous coursework for all students, from middle school through graduate school. This interest persists to this day as researchers currently investigate access to dual enrollment programs, career academies, and the International

Baccalaureate and Advanced Placement Programs in the state of Florida, using extremely rich and detailed data warehoused in the State of Florida's Department of Education.

The National Science Foundation's STEM Talent Expansion Program, or STEP Type II grant, supported the effort that is documented in this book. For this project data collection at each campus occurred bi-annually over a three year period. To ensure rigor among the researchers and through time, AAREA researchers received yearly data collection and field method trainings. All efforts to maintain confidentiality of the research participants were taken in the field and during analysis. This research included collaboration among the research team and research sites and ongoing support from our sponsor, in this case, the National Science Foundation.

Data and Methods

The research study reported in this volume focuses on the state of Florida and its postsecondary public education sector. We examine the four largest civil and electrical/computing engineering programs in the state. Florida is an important state to consider in this context because the state's racial and ethnic demographics are similar to those of the entire country including the projection of a growing Hispanic workforce. Although it is useful to understand the preparation of undergraduates in elite private institutions, it is even more important to examine undergraduate education in public institutions because they serve the largest numbers of students nationally.

Florida PK-20 Education System

Engineering education in the state of Florida is defined by the Florida PK-20 (prekindergarten to bachelor's degree attainment) education system linking PK-12 education, community colleges, and universities to provide continuous pathways to engineering and other degrees. Six universities in the Florida State University System offer engineering. Five universities that house four engineering departments are included in this study.[4] The University of Florida (UF) in Gainesville is the state's flagship university and one of only three universities founded before 1956. UF boasts the largest engineering program in the system. Florida State University (FSU) and Florida Agricultural and Mechanical University (FAMU), both in Tallahassee, are also older universities. FAMU is the only historically black college or university (HBCU) in the state system. These two universities share the FAMU-FSU College of Engineering, located equidistant from both campuses. Each university has its own faculty, but shares courses and facilities.

Along with UF and FSU, the University of South Florida (USF), with the main campus located in Tampa, is the state's third Research I Carnegie classification university. Founded in 1956, USF was the first of eight state universities founded in the latter half of the twentieth century and designated as a Hispanic serving institution with a population of 12.9% undergraduate Hispanic students in 2007. Florida International University (FIU) in Miami is unique among American universities, boasting a majority-minority student population, including 63.8% undergraduate Hispanic students in 2007. These four public university engineering programs were selected to provide diverse perspectives on the culture and climate of engineering education in the state of Florida.

The research agenda utilizes the unique institutional arrangements of Florida by examining community colleges functioning as feeders to the selected universities. Policies unique to Florida encourage community college attendance and have established community colleges as appropriate pathways to upper-level studies and bachelor's degree attainment. Over half of juniors and seniors in Florida public universities attend a community college for at least one course or earn an associate's degree (Venezia & Finney, 2005). All associate degree recipients are guaranteed entry into a Florida public university.

Florida's 2 + 2 programs include two years of community college study that lead directly to two years of junior- and senior-level coursework at one of Florida's state universities. These programs utilize a common prerequisite policy requiring any discipline providing a bachelor's degree to have common requirements taken at any Florida community college.[5] Each course offered in community colleges, public universities, and vocational centers is included in the common course numbering system, guaranteeing the course will be accepted and awarded the same number of credits at any other institution in the system.

Within this unique arrangement, the Florida Community College System offers key pre-engineering courses, including Calculus I and II and Physics I and II, at 28 community colleges and 61 campuses. No Florida resident lives more than 50 miles from the closest community college. For many students, these institutions are vital steps on the pathways toward bachelor's degree completion, including completion of the engineering degree. Many community college students generally transfer to the closest university after two years. Still, community colleges located near universities often serve as an affordable and flexible way to gain admittance to the local university; thus a purpose of this study is to examine the impact of community colleges' preparation of students for the culture and climate of the engineering programs to which they transfer.

Each of the four engineering programs included in this study (FAMU-FSU, FIU, UF, and USF) draws heavily from one or more local community colleges. Santa Fe Community College (SFCC) in Gainesville is located only minutes from UF and enrolls over 16,500 students. Tallahassee Community College (TCC) enrolls over 13,000 students from Leon, Gadsden and Wakulla counties (Florida Department of Education, 2007). TCC student graduates attend both FSU and FAMU, including students who transfer to the joint engineering program. The Partners in TCC program guarantees Tallahassee Community College associate's degree recipients who reach certain requirements admittance into FSU. Hillsborough Community College (HCC) enrolls over 38,000 students and consists of eight campuses located throughout Hillsborough County, the second largest county in Florida. Three HCC campuses are included in this study: Brandon, Dale Mabry (Tampa), and Plant City. With over 66,000 students, Miami-Dade College is the largest college in the United States; it boasts seven campuses in the Miami-Dade County area, and is the primary feeder institution into FIU (FLDOE, 2007). Miami-Dade College North and Kendall campuses are included in this study.

Instrument Development

Qualitative interview protocols, or guides, were designed by a team of researchers from I/O psychology, sociology, anthropology, and measurement. These protocols were developed following an extensive literature review which determined previously identified variables proven to be instrumental in STEM education for women and minorities. Different interview protocols were created for faculty, students, administrators, and staff as well as a focus group instrument for students. Pilot testing was conducted with engineering students outside of the sample to assure that the questions were clear and understandable to engineering students, and that the questions were correctly designed to elicit comprehensive responses.

The development of the two survey instruments (students and faculty) followed similar steps. First, we reviewed the literature on culture and climate in relation to organizational success. Most of this work was informed by I/O survey instruments developed to measure organizational values and found to have adequate reliability and validity. Items from these instruments were originally designed for work settings and thus were modified to suit an educational institution setting for the purposes of this research. From the review of these instruments, a pool of items was generated with a result of 119 items for the student survey and 40 for the faculty survey. The student survey was then piloted with undergraduate psychology students who did not participate in the final survey.

Site Visits and Data Collection

Throughout this three year project, our research team interviewed and surveyed students, faculty, administrators, and staff at participating universities and community colleges. We selected civil and electrical/computer engineering programs in large part because they enroll sufficient numbers of students to constitute a meaningful set for analysis of responses to student and faculty surveys. After receiving approval to conduct research from USF's Institutional Review Board, we used a variety of strategies to recruit participants. The principal investigator, Dr. Kathryn Borman, made the first contact, sending an introductory email describing the nature of study and requesting assistance with recruitment of participants. A series of communications with prospective participants followed. Where necessary, a member of the research team with ties to institutional "gate-keepers" established rapport and familiarized them with the site visit strategy that was a part of our multi-phase research plan.

During the first visit, the research team conducted face-to-face interviews with undergraduate engineering majors along with faculty, administrators, and staff. Students were selected based on their declared major in engineering, their status as a pre-engineering major, or undecided but likely to pursue engineering. While the sample target for students was civil and electrical engineering we did not exclude engineering students familiar with those departments, nor did we exclude pre-engineering students as they have not yet declared a concentration. Students were recruited via snowball sampling and through announcements made in engineering and pre-engineering courses and with the assistance of a student liaison assigned at each school we visited

Faculty, administrators, and staff were selected based on their knowledge of the departmental culture, specifically the goals and values of the department. Faculty interviews include tenure-track and tenured professors, assistant, associate, and full professors as well as adjunct part-time instructors. Administrators include deans, associate and assistant deans, and 18 staff interviews with academic advisors and office assistants who interact with students on a daily basis. Interview participants signed informed consents and their participation was voluntary. All interviews were audio recorded for transcription and analysis and strict confidentiality was ensured during and after data collection. Visits typically lasted five to ten days depending on access to students and other participants.

The student survey was administered over two terms. Before participating, we reminded participants of the purpose of the study, told participants the information they provided would be kept anonymous, and advised them they were free to decline participation at any time without penalty. Following administration of the student surveys, faculty surveys were administered in a similar format. Compared with students, the response rate for faculty was very low. The team made a conscientious effort to request faculty participation. Even with this strategy, only 24% of faculty ($N = 156$) responded compared to 80% of the students ($N = 1407$).

During the course of the study we earned the trust of most participants and, in fact, found that particular faculty and administrators embraced the purpose of the research. During years two and three of the study, our access to laboratories and to classes improved and researchers observed classes in session. Generally, student focus groups were intended to consist of members of one race and/or gender group. However some groups were combined due to difficulty recruiting students. Researchers also conducted additional face-to-face individual interviews with students, faculty, administrators, and staff during later visits.

Qualitative Coding and Analysis

Interview and focus group transcripts were coded in an Atlas-ti database using a codebook developed for the entire study. Atlas-ti is a qualitative software device that organizes transcripts and allows researchers to sort interview transcripts by codes. Codes are words or phrases that correspond to key ideas integral to the research, for example, if an interviewee were to discuss their experiences as a woman or minority a researcher would assign that segment of text to the code "experiences of women and minorities." Atlas-ti can then query all documents for segments of text assigned to this code. Depending on the focus of the chapter in question, members of the analysis team coded segments of transcripts corresponding to specific topics and with the central topic of a given chapter. For example, chapter four examines student preparation, development of interest in engineering, and classroom pedagogy. Analysts coded interviews and focus group data to "background" and "curriculum/pedagogy" codes. The background code included sub-codes for "engineering interest" and "preparation." The curriculum/pedagogy code included sub-codes "course work" and "professors." Chapter authors utilized Atlas-ti to select text segments included in these sub-codes and then analyzed them to uncover themes, or discernable patterns in the interviews. For example, if all students reported dissatisfaction with PowerPoint lectures that would constitute a theme for pedagogy. We employ a mixed methods research design in which qualitative analysis occurs first, with subsequent survey analyses occurring at the data interpretation stage (Leech & Onwuegbuzie, 2009). Based on the themes uncovered by the qualitative component, we identified items from the survey that were closely related to facilitate comparison of results from both analyses.

We used information obtained from analysis of selected survey items to complement and clarify themes derived from the qualitative analysis.

The research team consisted of two working groups: one for the collection and analysis of qualitative data based on observations, interviews, and focus groups with students, faculty, staff and administrators; the other for quantitative data collection and analyses based on the results of a newly developed survey. While this arrangement worked well during the data collection phase of the study, during the analysis and writing process we subsequently met as a group to pull together ideas and concepts held in common. Each chapter was a collaborative effort with the listed authors taking the lead and the editors and other members of the research team providing input and editing assistance.

Quantitative Analysis

Survey analysis was conducted after site visits. Data were first "cleaned" to remove invalid data entries and missing data. Descriptive analysis computed item means and standard deviations by gender and ethnicity for the sample. In addition, independent sample t-tests and chi-square tests by gender and ethnicity were conducted. These tests determined the extent to which men and women as well as White, Black, and Hispanic students differed from each other in their survey responses. We conducted separate descriptive statistics both for lower division (freshman and sophomore) and upper division (junior and senior) students. This distinction is important because we expect perceptions among these students to differ. Freshmen take a wide variety of courses, most from outside the engineering college, while seniors take courses primarily from a specific department. Moreover, seniors have more experience with the department than freshmen are likely to have and as a result their perceptions may be more accurate or reliable. Additional analyses established the reliability and validity of individual survey items. Further analyses including factor analysis are explained in more detail in chapter five.

Limitations

We encountered several unforeseen limitations. First, while researchers were committed to mixed methodology in design, implementation and analysis, it is difficult to balance qualitative and quantitative epistemologies and methods. We have striven to use data derived from both methods simultaneously to strengthen findings, but it has proved challenging to align multi-disciplinary understandings of culture and climate. As a result, our understanding of departmental culture relies on anthropological and I/O psychology understandings of culture. Our understanding of climate relies on theoretical conceptualizations of fit derived from I/O psychology. Second, we were interested in determining the extent to which students and faculty differed in their perceptions of various aspects of program culture and climate; however, the relatively low number of participants in the faculty survey ($N=156$) compared with participating students ($N=1407$) made it difficult to evaluate responses. Some of this disparity was due to the greater number of students compared to faculty, but most was due to low faculty response rates.

Third, it was difficult to contrast faculty and students' perceptions. Even though scores from student and faculty surveys were reliable, comparing responses to each survey was not as valid a measure of differences between student and faculty perspectives as we had hoped. Fourth, inconsistency in statements about institutional goals and what faculty and students perceive as important made it difficult to determine program culture. For example, a department or program's website may suggest that it values student support; nonetheless, students may mention lack of support services. Cognizant of the possibility that what people say or claim may not match espoused institutional values and actual practices, we use multiple data sources to strengthen our analysis. In addition to student and faculty surveys, the data from student interviews; faculty, administrator, and staff interviews; classroom observations; and archival data were triangulated. Concerns about confidentiality limited our ability to examine race and gender effects among faculty. Minority and female faculty and administrators make up a small number of all engineering faculty at each university. For this reason, statements by faculty, administrators, and staff are not identified by their race or gender. We are aware that the gender, race, and nationality of these institutional gatekeepers inform their opinions and experiences. This approach in no way undermines this fact or assumes that faculty, administrators, and staff speak with a single voice. We have made a decision to respect the identities of these valued contributors to this project by identifying them by their university and position.

Finally, meanings of terms including "department," "program," and "major" vary from institution to institution and it is necessary to define these terms to participants. For the purpose of the present study, we used the term *department* to refer to a division within a college devoted to a particular academic discipline through which a student is working to fulfill the requirements of the bachelor's degree.

Book Overview

This book examines six key components in pathways to an engineering degree. Chapter two describes Florida as a microcosm of the national picture.

We analyze four Florida state university engineering programs (comprised of five universities) with respect to location, demographic makeup, campus ecology, resources/infrastructure, and overall campus climate. The underlying research question is, "What contextual factors (culture and climate) in STEM affect students?" Chapter two has three goals: first, to introduce the research sites using observations of campus spaces and places to contextualize interview and survey data; second, to present holistic, mixed-method research along with ethnographic methods illuminating the cases; and third, and most importantly, to delineate how Florida is representative of national trends in the underrepresentation of female and minority recruitment and retention in engineering and to expand upon the concept of program efficacy. This chapter provides the groundwork for subsequent chapters focused on the culture and climate of undergraduate engineering programs in Florida.

Chapters three to five revisit the Florida State University System engineering programs and institutions presented in brief ethnographies in chapter two and build upon analyses of culture, climate, and curriculum to identify primary and secondary obstacles students face and, most importantly, the strategies used by students, faculty, staff, and administrators to cope with and overcome them.

Chapter three looks at "switchers," or students who have changed majors and switched from engineering into another field. Using interviews with "switchers" and administrators who work with these students, chapter three illuminates how women and minority students make the decision to leave engineering. This chapter seeks to present the most salient obstacles to student success to contextualize and ground the recommendations found throughout the book. The chapter outlines factors influencing attrition and persistence along engineering pathways among women and minority civil engineering and electrical/computing engineering students.

Chapter four uses qualitative and quantitative data to understand how the culture of undergraduate engineering programs at the five Florida universities is related to the retention of women and minority students in these programs. In particular, this chapter outlines what departmental values were reported by faculty, staff, and administrators and whether or not the reported values—or espoused values—were, in fact, translating to the students within the department. Tensions between faculty teaching and research obligations are well known in academia and this chapter addresses student interactions with institutional support and how they must navigate their programs when they do not feel supported by faculty. Specifically, women and minority students feel less supported and as a result rely more heavily on one another. Recommendations regarding formal support programs, continued and enhanced support of student organizations and programs to involve undergraduates in research are outlined as well.

Chapter five addresses the climate of engineering departments, defined as student perceptions of "what it is like" to be in the department with respect to practices, policies, procedures, routines, and rewards. Quantitative survey data suggest there are three aspects of department climate that are predictive of a student's intent to leave an engineering program: institutional support, personal agency and peer support, and perception of fit. Interestingly, quantitative analyses showed no statistically significant difference for race or gender among these factors, but qualitative analyses yield specific themes for women and minorities. Overall, students report that women and under-represented minorities are not treated differently; however, their narratives suggest they indeed do experience departmental climate differently than their white male counterparts. This chapter addresses this apparent contradiction, and conclusions highlight the importance of mixed-method research and the subtlety of a "chilly" campus for women and minorities.

Students' narratives of their classroom experiences, professors' instructional approaches, including both student and faculty survey responses, are integrated with data from class observations, faculty interviews, and surveys in chapter six. How students experience classroom instruction is the issue analyzed here in addition to unraveling the multiple reasons students chose engineering. These are explored within the context of family background (e.g. parents' occupational careers, first generation in college, etc.) and educational preparedness (e.g. community college, high school courses). The chapter concludes that building an early and enduring interest in engineering is critical, especially for women and minorities.

An important aspect of this study concerns two year institutions or community colleges. Many students who eventually pursue engineering begin their studies in community colleges. Chapter seven examines early college science pathways and the challenges students may encounter as they transition from a community college to a four-year university. By examining focus group, interview, and workshop data from both community colleges as well as four-year universities, we theorize what practices and policies are most conducive to a successful transition from a community college to an engineering program at a four-year institution. In addition, we identify which practices and policies are most problematic for student success.

The conclusion, chapter eight, takes up the patterns, themes, and relationships in the data presented in the preceding individual chapters, both quantitative and qualitative, and macro (climate and culture data) and micro (student, faculty, and administrator data). It also identifies and analyzes inconsistencies, contradictions, and conflicting patterns in the data. For example, some students appreciate small classes; others are intimidated by them. This chapter outlines the potential improvements to programs as iterated by engineering students as well as recommendations that emerged through data analysis. The concluding chapter aims to summarize what we argue could be done to ensure more students—particularly women and under-represented minorities- become engineers.

The audience for this book is broad; engineering faculty, administrators, and students, as well as policy makers in education and science at various levels of government will benefit from reading this book and understanding its implications. By writing this book, we hope to impact science education and policy at many levels: secondary, undergraduate and graduate, as well as locally, regionally, and nationally. Specifically, our comparisons of engineering programs at the institutional level provide a revealing look at effective strategies used by institutions, faculty and administrators, and especially by the students attending them that are most efficacious for surviving and thriving in higher education.

Notes

1. Throughout the following chapters, the term Black is used interchangeably with the term African American.
2. See, for example, the highly contested book the *Bell Curve* by Richard J. Hernstein and Charles Murray first published in 1996.
3. The engineering programs at the University of Central Florida (Orlando) and Florida Atlantic University (Boca Raton) are not included in this study.
4. See note 2.
5. The FLDOE allows students to construct a personal education plan using common prerequisites manual: http://www.facts.org/PreCoreq_SW/PreCoreq2006/index2.html/.

Chapter Two
Producing STEM Graduates in Florida: Understanding the Florida Context

Bridget A. Cotner, Cassandra Workman Whaler, and Will Tyson

Introduction

After making the case for studying STEM production in a national context in chapter one, chapter two narrows the focus to Florida, considering the state as a microcosm of issues confronting students pursuing STEM degrees across the country. Specifically, our discussion includes descriptions of the locations, demographic makeup, resources, and infrastructure, and overall campus ecology of the four engineering programs we focused on. The use of a mixed methods research approach allows this chapter to meet three goals. The first is to frame Florida as a unique field site that is also representative of national trends in engineering, including recruitment and retention of underrepresented women and minority students. A discussion of program efficacy provides information on how it is calculated and defined. The second goal is to familiarize the reader with ethnographic methods. Finally, we will apply our ethnographic methods to introduce the reader to the research sites, using observational and anecdotal information to contextualize interview and survey data presented here and in subsequent chapters. In subsequent chapters, we more closely examine culture and climate as they affect student fit and student retention in engineering undergraduate programs in the state.

The National Context

An overview of the national context is embedded in the following section to familiarize the reader within current conditions affecting engineering students. We also address the critical need for female and minority undergraduate engineering students to understand shortfalls in the number of individuals who enter the U.S. labor market. This section concludes with a discussion of Florida's position vis-à-vis national trends in engineering higher education.

STEM Production in Florida State University System

Using information obtained from secondary sources and from related literature, this section describes STEM production in the Florida State University System (SUS), specifically comparing and contrasting Florida four-year public universities with national norms. These comparisons underscore the fact that the Florida SUS is a rich field site representing a microcosm of U.S. public universities.

Racial Diversity

As shown in figure 2.1, the number of Florida bachelor's degree recipients has almost doubled from 26,403 in 1990–1991 to 47,326 in 2006–2007. This increase in degree attainment can be attributed to several factors. First, propelled by increases in state aid for the brightest students, a growing population, and other factors, seven universities increased their enrollments by at least 50% between 1990–1991 and 2006–2007. Enrollment at the University of Central Florida (UCF) more than doubled over this time period making UCF one of the 10 largest universities in the country

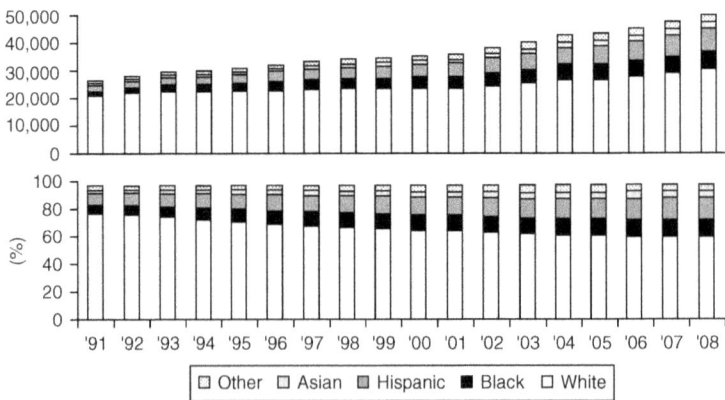

Figure 2.1 Florida Bachelor's Degree Recipients
Source: Adapted by the authors from the Florida Board of Governors Interactive University Database.

along with the University of South Florida and the University of Florida. Second, the state of Florida opened Florida Gulf Coast University (Fort Myers) in 1997 and New College of Florida (Sarasota) in 2001. Third, Black student enrollment doubled and Hispanic student enrollment tripled from 1990–1991 to 2006–2007. By comparison, overall enrollment increased by 22%.

White graduates made up almost 80% of all graduates through the early 1990s; these numbers have declined to under 70% by 1998 and currently remain slightly more than 60%. Decreases among White graduates are complemented by substantial increases among Black and Hispanic graduates. Black students made up only 6% of all SUS graduates in 1991, increasing to 12% in 2000 hovering between 12 and 13% throughout 2007. Hispanics made up 9.1% of SUS graduates in 1991 and have increased steadily to 16.5% in 2007.

Trends in Florida engineering degree recipients are shown in figure 2.2. White engineers make up around 55% of engineering graduates since 2000 down from over 70% in 1991. Black engineering degree recipients increased from 3.8% of all graduates in 1991 to a high of 14.3% in 2000 equal to Hispanic engineering degree recipients who increased from 11.2% to 14.4% over the same time period. Since 2000, Black engineering degree attainment has declined to 10.0% of all those receiving the degree in the state while Hispanic degree attainment has increased to 19.2% of the engineering degree recipients. Asian students are overrepresented in engineering, making up between 8 and 10% of all engineering graduates over the last 10 years compared to between 2.2% and 4.6% of all graduates from

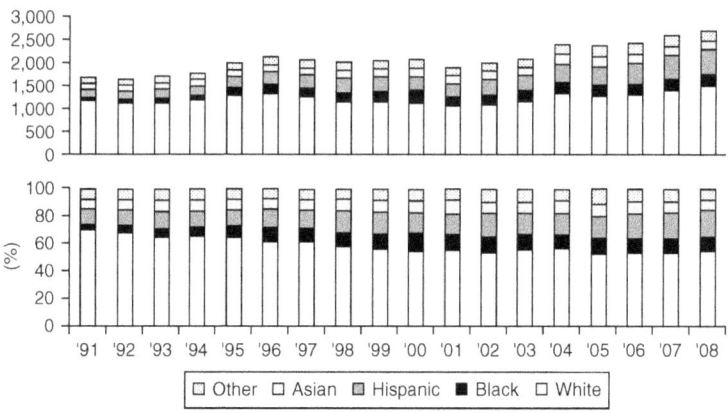

Figure 2.2 Florida Engineering Degrees

Source: Adapted by the authors from the Florida Board of Governors Interactive University Database.

1990–1991 to 2006–2007. Although there has been some progress, the low levels of engineering degree attainment among Blacks and Hispanics in Florida and nationally in comparison to White students heightens the importance of our research into engineering careers.

Gender

National and state data show the emergence of women in higher education through the turn of the present century. Until the 1970s, women were a distinct minority in higher education; recent analyses of national census data indicate that women now surpass men in attainment of undergraduate college degrees as reflected in returns to gains in the standard of living for women that have vastly improved in comparison with men over the period from 1964 to 2002 (Buchmann & DiPrete, 2006). Women now compose 53 to 57% of Florida university enrollees from 1990 to 2007 and have increased their share of all degrees from 56% to 60% over that time period. By comparison, women increased from 17.6% of all engineering enrollees in 1991 to almost 21% from 1998 to 2001. From 2001 to 2007, percentages have fallen to under 18%. Women's share of engineering graduates grew from 16% in 1990–1991 to 23.7% in 2001–2002, but fell to 19.5% in 2006–2007. Overall, Florida public institutions exhibit patterns of change in higher education that model national changes in public higher education.

Program Efficacy

As explained in chapter one, this book uses *program efficacy* instead of *retention rates* to describe the progress of underrepresented students in a specific higher education environment without respect to the micro-level or student-level idiosyncrasies that may interfere with bachelor's degree attainment. Program efficacy is a normative term referring to programs that have *high value,* in other words, programs in which women, African-Americans, and/or Hispanic students exhibit greater progress toward degree attainment than underrepresented students in other programs and within the general student population across the institution. In *low efficacy* programs, these underrepresented students do not progress toward degree attainment at a comparable rate to peers outside the program. This method of calculating program efficacy compares the change in racial and gender composition on program and institutional levels at three stages of higher education pathways:

1. Lower-level undergraduate—Students classified as first-year or sophomore students, typically students in their first two years at the

university. Engineering lower-level undergraduates take core engineering courses along with courses in physics and calculus.
2. Upper-level undergraduates—Students classified as juniors or seniors. These students have completed courses at the lower-level and are on track to earn an engineering degree. These students take required courses in the major as well as topical elective courses.
3. Bachelor's degree recipients—Upper-level undergraduates who completed a bachelor's degree in an engineering field.

To ensure a balanced comparison, this method uses data only for students whose first enrollment was at the university or First Time in College (FTIC) students; a small percentage of these students include transfers from community colleges. With the exception of a small number of students with enough credits to enter as a junior, upper-level FTIC undergraduates were at one time classified as lower-level undergraduate and bachelor's degree recipients were at one time classified as upper-level undergraduates. However, we cannot assume that all lower-level undergraduates who are not upper-level undergraduates in two years are no longer on the pathway to degree attainment since many students attend college from four to six years on average.

Our measure of program efficacy compares the differences in racial composition of FTIC students at each level during a school year. For example, if women make up 50 % of lower-level undergraduates enrolled at a university in 2000–2001 as well as 55% of upper-level undergraduates, the retention ratio is 55%/50 % or 1.10. This measure simply indicates the gender composition of upper-level students comprises a cohort of 10% more women in comparison to the lower-level composition. If women make up 60% of degree recipients compared to 55% of upper-level undergraduates, the persistence ratio is 60%/55 % or 1.09.

At this hypothetical university we should expect women to make up a greater percentage of engineering students at each level if the composition of the engineering program after initial enrollment is comparable with that of the entire institution. Let us assume that 20% of lower-level undergraduates enrolled as engineering majors are women. Given that women make up 10% more of upper-level undergraduates than lower-level undergraduates, we would expect the engineering program to be comprised of 20%*1.10 = 22 % of women. Assuming 22% of upper-level undergraduates are women, we would expect 22%*1.09 = 24% of engineering bachelor's degree recipients to be women. This study regards programs as highly efficacious in retention if the retention ratio of upper-level engineers to lower-level engineers is 0.10 higher than the retention ratio throughout the university. In this example above, the engineering program would have high efficacy retention if women make up 1.20 times more of upper-level engineering than lower-level engineers.

Programs are highly efficacious in persistence if the ratio of degree recipients to upper-level students is 0.10 higher than expected. The engineering program in question would be highly efficacious if women make up 1.19 times more degree recipients than upper-level engineers. In contrast, programs that are considered low efficacy in either persistence or retention have a 0.10% lower ratio than expected. In this example, an engineering program in which the ratio between upper-level women and lower-level women is 1 would be considered low efficacy retention. This engineering program would be considered low efficacy persistence with a degree recipient to upper-level ratio of 0.99, 0.10 less than the expected ratio of 1.09. This means if women in the university as a whole outperform men but only perform the same as men in engineering, the engineering program is low in efficacy.

This measure compares retention and persistence within engineering programs using the FTIC students enrolled in engineering and in the university as a whole rather than showing susceptibility to selection biases governing the race/gender composition of incoming cohorts of engineering programs and the university as a whole. This measure also captures the relative progress of race and gender groups without crediting or punishing an engineering program for following trends experienced in other areas of the university. For example, if in a particular university, from 1998 to 2007, women made up slightly more degree recipients than upper-level undergraduates and slightly more upper-level undergraduates than lower-level undergraduates. A highly efficacious engineering program would be above that benchmark.

Methods

This chapter employs quantitative and qualitative methods to understand the institutional context in which learning and engineering degree attainment occur in order to explain some of these trends. Each institutional profile includes demographic information as well as program efficacy over time and ethnographic descriptions of campus life. Each profile includes (1) history, (2) student/faculty demographics and change over time, (3) program efficacy over time, and (4) campus ecology of the college of engineering, that is, the university infrastructure and the physical environment and how students utilize resources. As described below, researchers utilized anthropological or ethnographic methods to discern campus ecology. Taken together, these background and ethnographic data set the stage for the research conducted for our study.

Ethnographic Methods

Low levels of engineering degree attainment by Blacks and women accentuate the need to understand the context of engineering programs. Ethnographic methods are well suited such work (LeCompte & Schensul, 1999). Originating in anthropology, ethnography is designed to gain an "emic," or insiders' understanding of cultural phenomena. To achieve an emic perspective, ethnographers employ a variety of methods including participant observation and speaking with research participants. While the hallmarks of ethnography are long term contact and immersion in a research site, research has demonstrated that the use of rapid ethnographic methods also produces valid results. Regardless of duration, the use of multiple qualitative and ethnographic methods can present a more holistic view of a culture, in this case the culture of academic departments such as civil and electrical engineering. Specifically, in this study we sought to understand what contextual factors enhance student retention and student fit at each participating university.

Five Florida Public Universities: Profiles of Colleges of Engineering

The profiles of the five universities and four colleges of engineering that make up the setting for this study portray institutions with different purposes and goals: Florida Agricultural and Mechanical University (FAMU)/Florida State University (FSU) College of Engineering, Florida International University (FIU) College of Engineering, University of Florida (UF) College of Engineering, and University of South Florida (USF) College of Engineering comprise the institutions included in this research. Figure 2.3 shows the location of the five universities in the study. The five universities of interest in this study share the task of educating a diverse Florida student body and they each serve distinctly different student populations.

Site 1: Florida Agricultural and Mechanical University (FAMU) and Florida State University (FSU) Joint Engineering Program

History and Mission

Florida Agricultural and Mechanical University, also known as Florida A&M or FAMU is a recognized historically Black college or university (HBCU) in Tallahassee, Florida. Founded in 1887 as the State Normal College for Colored Students, the name changed to Florida Agricultural and Mechanical University for Negroes in 1909 and finally to the current

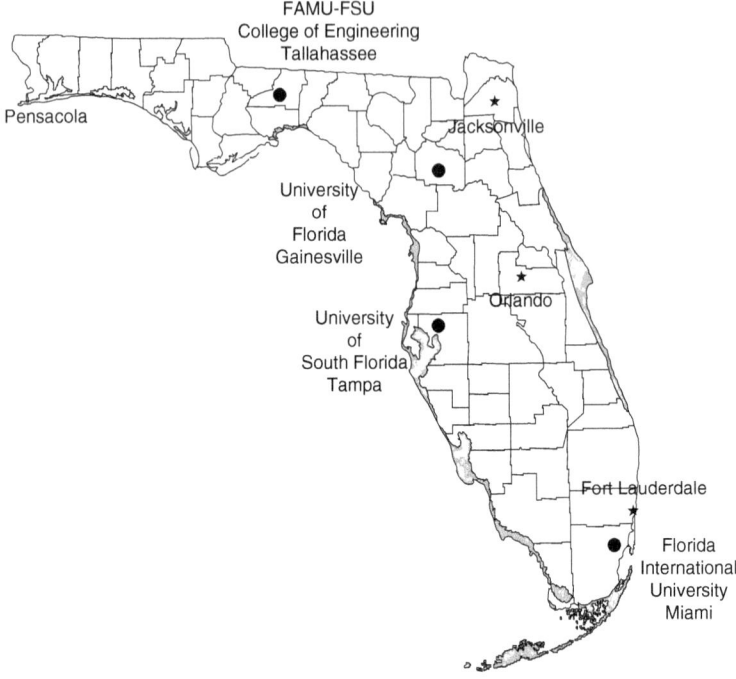

Figure 2.3 Five Florida Public Universities
Source: Adapted by the authors from the Florida Board of Governors Interactive University Database.

Florida Agricultural and Mechanical University in 1953. FAMU ranks 15th for Tier 1 universities of the U.S. News & World Report ranking of HBCU's for 2009. FAMU opened with an initial cohort of fifteen students and now enrolls more than 11,587 students. Created in 1987, FAMU's College of Engineering is a joint program with Florida State University (FSU) and is unique in that regard.

Originally established in 1851, FSU is the oldest state university in Florida. Florida State University (FSU) began as one of two universities founded in the state of Florida (the other is the University of Florida (UF)). FSU was first called the Florida Technical Institute and later the West Florida Seminary serving young women. During the Civil War, the name was changed to the Florida Military and Collegiate Institute. The name was later changed to Florida State College. Later a school for women, Florida State College for Women (FSCW) was created and eventually became coeducational. At that time the institution acquired its current name, Florida State University, in 1947. The first African American student, however, was not admitted until almost 20 years later in 1962. Presently, the race and gender composition of FSU is comparable to the entire Florida SUS system.

Unlike other colleges of engineering in the state of Florida, FAMU's and FSU's College of Engineering was established in 1982 as a joint program serving both institutions. The joint engineering program is a structure that is unique in Florida and is sometimes challenging for administrators and faculty members who highlight the difficulties associated with serving two institutions. A Black administrator in the engineering program explained, "Everything that we do has to be in common because we serve the two universities so we have to find some common ground." The mission of the College is to provide an innovative academic program of excellence at both the undergraduate and graduate levels and to attract and graduate a greater number of minorities and women in professional engineering.

The FAMU-FSU College of Engineering enrolls only 12% of Florida engineering majors, down from 16% in the 1990s. Engineering enrollment reached 813 in 1998, but decreased to less than 500 by 2006. A more detailed look at Black engineering majors indicates that the decrease at FAMU is not due to Black engineers choosing other programs, but because of the statewide decline in Black engineering majors more generally. Both FAMU and FSU enrolled fewer Black engineering majors in 2007 than they did just 10 years before.

Students and Faculty

Although students matriculate and earn their degrees from a particular university, they share resources in the FAMU-FSU College of Engineering joint program. The joint program currently has 90 faculty members serving five academic engineering departments: Chemical & Biomedical, Civil & Environmental, Electrical & Computer, Industrial, and Mechanical. During the fall of 2006, 1,906 undergraduate students were enrolled in the college. Of these, 301 declared civil engineering and electrical/electronic engineering majors. Figures 2.4 and 2.5 show key elements of FAMU and FSU engineering enrollments from fall 1998 to fall 2007:

- Percent of engineering majors who are women
- Percent of engineering majors who are African-American
- Percent of engineering majors who are Hispanic
- Percent of all undergraduate students who are engineering majors

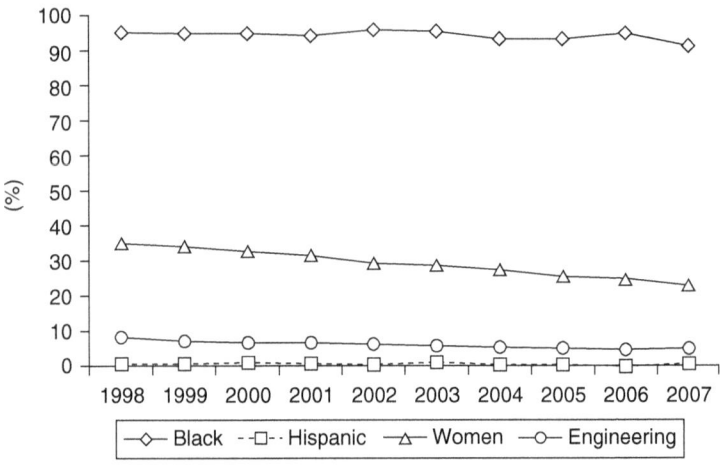

Figure 2.4 Florida A&M
Source: Adapted by the authors from the Florida Board of Governors Interactive University Database

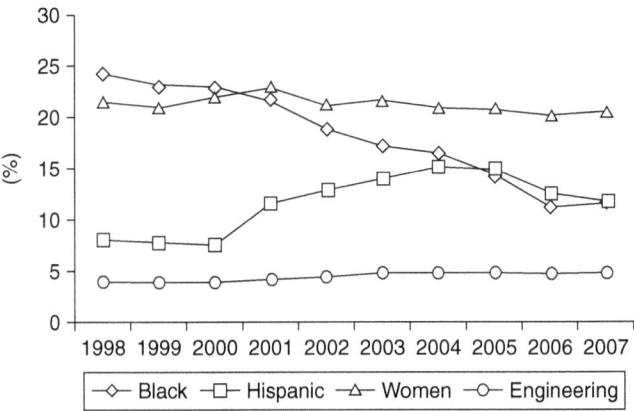

Figure 2.5 Florida State University
Source: Adapted by the authors from the Florida Board of Governors Interactive University Database.

As the only public HBCU in Florida, it is not surprising that a large majority of engineering majors are African-American. The decline in female engineering majors from 35% to 22% from 1998 to 2007 is dramatic. This trend mirrors national trends for a decline in women engineering majors (Seymour, 2001).

Hispanics make up less than one percent of engineering students, reflective of the small number of Hispanic students enrolled at FAMU.

There is an interesting story to tell at FSU. African-American engineering enrollment has decreased while Hispanic engineering enrollment has increased. In raw numbers, African American enrollment was stagnant with a sharp decline since 2005 while Hispanic engineering enrollment almost tripled. There was little change in female enrollment or engineering enrollment compared to total enrollment.

Program Efficacy

Because FSU and FAMU are joint programs but admit students and hire faculty separately, this chapter applies two different measures of program efficacy to compare retention and persistence in the joint engineering program to each campus. Retention and persistence among Black FAMU engineering students is comparable to that of the entire institution. From 1998 to 2007, the university retention ratio among Black students improved from 0.93 to 0.99, comparable to the engineering retention ratio over that time frame of between 0.98 and 1.02 among Black engineers. Persistence is similar ranging from 0.98 and 1.03 among all Black students and between 0.97 and 1.03 among Black engineers. Efficacy among Hispanic students at FAMU is not reported because there are very few Hispanic students enrolled at FAMU.

During more recent years FAMU-FSU maintained high efficacy with respect to FAMU women. Between 1998 and 2007, the retention ratio among all FAMU women improved from 1.00 to 1.09 as female enrollment stayed constant and women comprised a higher percentage of upper-level undergraduates. Despite declining enrollment among FAMU engineering students from 35% in 1998 to 24% in 2007, women made up 34% of upper-level engineers in 1998 remaining constant until falling to 29% in 2007. The retention ratio in 1998 was 0.97, but increased to a high of 1.53 in 2004. FAMU women in FAMU-FSU engineering had highly efficacious enrollments from 1999 to 2006 with the exception of 2001.

FAMU-FSU also promotes persistence among women, but not as consistently as retention. Overall, women make up a slightly larger percentage of degree recipients than upper-level undergraduates, but the persistence ratio decreased from 1.05 in 1998–1999 and 1.11 in 2003–2004 to a low of 0.99 in 2006–2007. FAMU women enrolled in FAMU-FSU had a high efficacy persistence ratio in comparison from 2000–2001 to 2002–2003 and in 2004–2005, but low efficacy in 2003–2004 and 2005–2006, a sign that trends in degree attainment among FAMU women at FAMU-FSU are in flux.

During the fall of 2006, FAMU enrolled half as many female engineering students as in fall 1998. This suggests that FAMU's recruitment of female engineering students over time has declined. Furthermore, FAMU's enrollment is becoming an increasingly smaller portion of the engineering program as a whole. In 1998, FAMU students comprised 46% of all lower-level engineering students. By 2007, FAMU students were only 24% of lower-level engineering students. From 1998 to 2007, FAMU upper-level engineering students decreased from 424 to 158, falling from 37% of upper-level engineers to only 11% in 2007. Degree attainment has fallen as well from 40% in 1999–2000 to only 13% in 2006–2007.

Overall, FAMU-FSU is a highly efficacious program for FSU students. Enrollment among Black students has declined from 13% to 9%, but Black students consistently make up around 12% of upper-level undergraduates and from 10 to 12% of degree recipients from 1998 to 1999 to 2006–2007. These same enrollment declines can be found in FAMU-FSU engineering. Black engineers made up 29% of all lower-level FSU engineers in 1998 but decreased to 10% in 2006. Upper-level, junior and senior Black student engineers decreased from 21% of all upper-level engineers in 1998 to 11% in 2007, mirroring degree attainment in the same school years. Black retention ratios stay below zero, but are only low in efficacy in 1998 and 2002. Despite the smaller percentage of Black FSU engineers among upper-level FAMU-FSU engineers, the persistence ratio varied greatly. Black FSU engineers show high efficacy in persistence in 1998 and from 2001 to 2005 with the exception of 2003.

Growth in Hispanic enrollment at FSU corresponds with increased representation of FSU Hispanic engineers. As would be expected with such rapid growth, the ratio of upper- to lower-level undergraduates is below 1.00 as enrollment trends catch up. Persistence ratios are strong most years as Hispanics are better represented among degree recipients than all upper-level undergraduates. Retention and persistence ratios are fairly low until recent years. FSU Hispanic engineers are low efficacy in 1999 and high efficacy in 2000, 2004 and 2005. FSU Hispanic engineers show improvement in persistence as well, moving from low efficacy in 1999–2000 through 2001–2002 and 2004–2005, but high efficacy in 2006–2007.

FSU women show the greatest efficacy at FAMU-FSU, particularly in recent years. FSU women showed high efficacy in retention in 1999 due to a small dip in lower-level engineering enrollment and small increase in upper-level enrollment. High efficacy in retention returned from 2004 to 2007. FSU had low efficacy in persistence in 1999–2000 and 2000–2001, but showed high efficacy in 2001–2002, 2002–2003, and later in 2006–2007.

Resources and Infrastructure

The FAMU-FSU joint engineering program is housed apart from both FSU and FAMU's main campuses at Innovation Park. It is located almost equidistant from the two institutions, three miles from FAMU and four miles from FSU. The College is a state-of the-art facility, a modern three-storied building constructed from glass and steel. This facility is divided into two wings. On entering the glass doors of wing A, there is a main large foyer, the Atrium, between the pathways designed to enter and exit the building. Sunlight entering the Atrium from the glass doors and windows above provides good lighting for the large number of students who study there. The classrooms are well lit and equipped with technology. There is a food court in the first floor where a variety of hot foods, drinks, chips, and cookies are sold. Apart from students standing in line in the food court, most are engaged either in study groups or are studying alone.

A large glass case in which trophies, award certificates, and pictures of students and faculty are displayed runs along the left wall near the entrance. Outside the Career Office is a rack containing magazines and other bulletins about engineering jobs, salaries, and related information. In the other hallways, bulletin boards are posted with useful flyers, posters of student and faculty projects, conference presentations, and scholarship opportunities.

The Atrium is designed primarily as a study area with chairs, desks and tables of all sizes so engineering plans can be laid out. There are also study carrels with computers and chairs along one of the walls as well as empty carrels to plug in laptop computers. The Atrium continues into wing B as a long hallway with computer carrels and lounge chairs on the left where students study. At the end of this hallway is a small cafeteria with tables and chairs for students to eat meals. While a few students studied in this area, most of them were eating and chatting loudly. The building is kept up well from the outside: a well groomed lawn and a convenient bus stop are located just a few yards from the main entrance to the building. In front of the building are ample parking lots both for regular vehicles and for alternative means of transport such as motorbikes and bicycles.

Campus Climate and Cultural Differences

Students gather in groups working on projects at all times of the day. Peak use of the Atrium is between classes, held between 9 a.m. and 5 p.m. In the hallways, students study individually and in groups often eating and talking during this process. Faculty also seem very busy—either preparing for a class, in the lab, or in a meeting with students or fellow faculty. Support staff members were very friendly to the research team.

Students were particularly happy to "own" a building set aside for engineering majors—a place where they felt their academic needs were being met. An African American female noted during an interview:

> ...so this one building is ours for Engineering. That's what makes it great just to know that everybody here is really trying to do these same kind of things. And just knowing that you can go out... You could just walk out and see one person studying and you know that they're doing something in Engineering so if you have a question you're just like, "Hey, can you help me?" And by it being here in this one building and then the department you get to know the people. You can recognize their faces from class. And it's not so spread out where you have to go all the way over to the other side to get somebody to help you. You can just walk around and find somebody. So I'm really excited that I chose this school.

Because the building is located away from both campuses, some administrators feel more independent and able to undertake activities to benefit their programs. A Black male administrator described some advantages of this arrangement:

> I think that... being away from both campuses is a strength for us because we're kind of independent in a way you know. And so we're able to do things, to get away with stuff that you wouldn't be able to do on main campus because we're a unique situation. So, typically the administrations of both [institutions] said: as the College of Engineering you guys kind of do your own thing you know. So we're able to set policies and things that will benefit us as opposed to having to follow something on either campus. Our general philosophy of that is that we look at both of them and whatever is the most stringent of the two we try to use it here at the college.

Similarly, an administrator points out that at the organizational level, they tend to be neutral, trying to treat all students as equals regardless of which institution might have admitted them. Asked about things that the departments or the college does to make students feel they belong, he notes:

> Well ... you know that's a kind of a unique question for us because of the two universities, right. So you're never going to see really any orange or green or garnet and gold in the building, right. You're going to see a lot of blue and white because we don't... we don't try to promote either university in terms of the two colors or the mascots or whatever... So there's an identity but there's not an identity associated with the college. And I mean I think it's something you're going to need to talk to the students about to see what their feeling is on that topic. But I know at least at the administrative level we try to be as neutral as possible in terms of that type of thing.

As evinced through their interviews, students also feel that their separation from main campus is beneficial as they know that they can always find someone from class to help them and that they are surrounded by engineers. In terms of forming study groups or making friends, White male students in a focus group reveal that a student does not need to know another student to engage in a study group when needing help:

> A lot of times I find if I'm just staying on campus and I'm doing my homework and I see that some other people I know from class they're all still on campus and I see that they're getting their stuff done too then I'll purposely align myself with them because I know they like to get it done. Like I usually try and group up with people that are trying to get it done too or that I know from other classes before, you know that kid's smart you got an A in that other class; I'm gonna go ask him.

The Atrium is an area where students feel comfortable approaching one another for help and to form study groups. While they may feel separated from the main campus, students do see the benefit to being surrounded by classmates and other engineers.

Site 2: Florida International University (FIU)

History

Florida International University (FIU), the youngest institution included in this study, was founded in 1972 to serve the then-booming Miami, Florida population. FIU was established in 1965 to primarily provide bachelor's degree opportunities for women in south Florida who were completing community college. FIU is the only Hispanic-serving public university in Florida and one of 193 such institutions in the country. The first enrollment cohort was 5,667 students and as a female administrator in the college of engineering pointed out, "growth has been rampant." Today FIU has a total student enrollment of 38,614, with 29,695 undergraduates.

The College of Engineering has 2,198 students, with 476 civil engineering majors and 390 electrical engineering majors. In contrast to UF and FSU, Florida International University has increased its share of engineering enrollments in Florida, growing largely because the number of Hispanic engineering majors at FIU more than doubled from 665 in 2002 to 1,461 in 2007.

FIU began as an upper-level university designed to serve students transferring from Dade County Junior College (now Miami-Dade College), and offered only upper level courses for the first nine years. A female administrator from the college of engineering explained, "Most universities have a very broad base of freshmen...but we started off as an upside down pyramid [admitting in only juniors and seniors]. We still have more than 60% of our undergraduate students who transfer from a community college or from another four-year institution. So there's still a lot of transfer population." In 1981, first year students and sophomores were able to matriculate and in 1984 the university started doctoral programs. FIU is comprised of two campuses, the main campus, University Park Campus in Miami, and Biscayne Bay Campus which is north of the main campus. In addition, the College of Engineering, comprised of one large building with an outcropping of labs, is located approximately a mile away from other colleges and departments at the main campus.

Student Demographics

As noted above, FIU draws heavily from the surrounding geographical area for its enrollment of undergraduate students. A female student in a focus group observes, "FIU is mainly for Miami-Dade people...and [students from] Broward County." Females in the focus group commented on the large number of non-white students, "There's more Hispanic students. I think there's more Hispanic and Indian and maybe [Asian] students than, I don't know, white." FIU reflects national trends in enrollment reflecting larger percentages of male engineering students in comparison with female undergraduates in engineering. A female student in a focus group interview observes about the low numbers of female students, "Like it's completely out of balance. Just because it's engineering. Yeah that happens at every school." A White female administrator from the engineering program stated, "It depends on how you look at diversity because a lot of people will look at the fact that we are 55% Hispanic and go, 'oh that's not very diverse'. And as you start looking at that [you] realize that with that there's a lot of diversity because those students come from Brazil and Venezuela and Cuba. We could always do better but you know I think this is a very diverse campus." The diversity of the Hispanic population is tied to the connections Miami has to the Latin American community itself.

Perceptions of the race and gender composition of FIU Engineering are quite accurate. Figure 2.6 illustrates the high Hispanic enrollment in FIU engineering, unique to FIU as well as rising Hispanic enrollment in engineering which can be found statewide. Female enrollment in engineering at FIU has increased as well. Gains among Hispanic and female students comprise much of the increase in engineering enrollment as a percentage of total student enrollment.

Program Efficacy

Increases in engineering enrollment correspond to increased efficacy among women and Hispanic students. Retention ratios among Hispanic

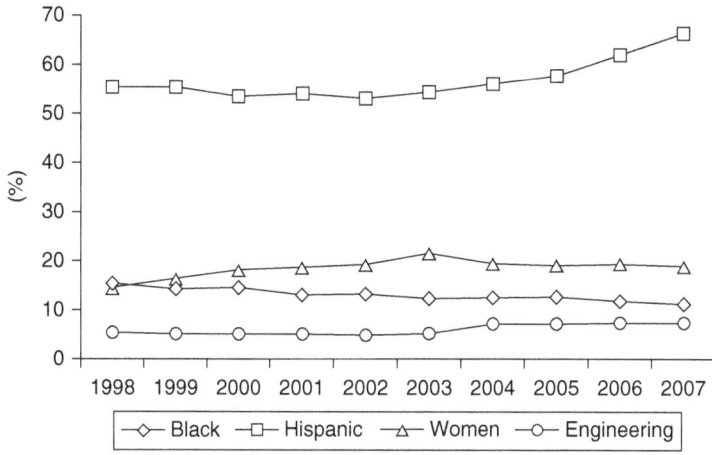

Figure 2.6 Florida International University
Source: Adapted by the authors from the Florida Board of Governors Interactive University Database.

students have decreased from 1.01 in 1998 to 0.93 in 2007 matched by an increase in persistence ratios from 0.93 in 1998–1999 to 0.99 in 2006–2007 showing that the Hispanic composition of degree recipients mirrors that of upper-level students. Engineering ratios show similar trends and were high enough from 1999 to 2001 to show high efficacy in retention. FIU's status as a Hispanic-serving institution may explain why this measure does not capture any major disparities in the racial composition at any level because Black engineers at FIU show declining efficacy.

Black FIU students show increasing retention ratios throughout the university from 0.88 in 1998 to 1.10 in 2007 compared to consistent persistence ratios at around 1.00 from 1998–1999 to 2005–2006 with a decline to 0.89 in 2006–2007. Black FIU engineers showed high retention ratios from 1998 to 2003 with the exception of 1999. This suggests that retention ratios among engineers were ahead of retention ratios on the rise among all Black FIU students. Retention ratios dropped below 1.00 for Black FIU engineers as campus-wide retention increased resulting in low efficacy from 2004 to 2007.

Black FIU students show persistence ratios lower than 1.10 for most years leading up to 2006–2007 due to declines in both upper-level engineers and degree recipients. Black FIU engineering students showed low efficacy in persistence from 1998–1999 to 2003–2004 with the exception of high efficacy in 1999–2000. A declining presence of Black engineering students at FIU is one of the key demographic shifts of the past ten years.

Women show increasing efficacy in response to increasing engineering enrollment. Retention ratios among FIU women engineers have been over 0.25 higher than the campus norm from 2003 to 2007. Likely due to such high upper-level enrollment, persistence ratios show little difference except for two years of low efficacy in persistence in 2001–2002 and 2003–2004 and one year of high efficacy in 2005–2006.

Resources and Infrastructure

During our first visit an undergraduate engineering female student asked us, "Have you seen the sign in the corner of the intersection? After a hurricane, it's like no paint." We were told that damage caused by Hurricane Andrew in 1992 had still not been repaired, symbolizing to these students the institution's lack of concern for the "forgotten corner" of FIU's sprawling campus. In contrast, other parts of FIU's campus are aesthetically very beautiful and the campus boasts new, architecturally innovative buildings. Sculptures throughout the main campus were donated from a philanthropist's private collection. As one walks through the main campus, gardens between the buildings become apparent. The student center on the main campus is bustling with students and there are students selling food or goods or, at the time of our visit, running for office. FIU's student center appears to be well utilized; students are everywhere studying, eating and talking. This contrasts sharply with the FIU College of Engineering. During our visit there in spring semester of 2007 the building had only recently reopened the "Panther Pit," an area for students to eat and to study.

The Panther Pit in the College of Engineering is designed to serve multiple purposes: To provide food and drink to students and faculty, to provide a place for students to meet to study or work on group projects together, and finally, to provide a place for students to relax and have fun. While the Panther Pit was open for business in the fall of 2006, there were still signs of construction and little evidence that it was well used by students for studying or relaxing. There were, at most, a couple handfuls of students using the Panther Pit at any given time. There was very little food offered besides snacks, and the only hot food available is hamburgers. Engineering students complain that prices are too high for the typical student. One female student explains, "No, the prices are not for a student. They charge you like five dollars, four dollars for just six inches with just one slice of ham and cheese you know. So it's not for students." Having access to food that was priced for the student budget was desired.

Despite these limitations, students are happy overall to have a place to go between classes. Many students told us how difficult it had been in the past to spend time between classes or find someplace to study. Unlike the atrium in the FAMU-FSU complex, there are no places for students to gather to work together in the FIU engineering building. One Black male student notes that engineering students find it difficult to locate a place to study, "Usually for us...I see a lot of the times it's just in the Panther Pit or any of the hallways that have chairs." Upon our return in 2007, the Panther Pit had grown in popularity and use. One female student comments, "And now, it's 100% right." The food options had expanded and a range of prices are available so that every student could afford to buy something to eat.

On higher floors there are study carrels with students using them. However, some students still find the carrels insufficient with one engineering undergraduate student saying his good grades were primarily related to his friendship with a number of graduate students together with whom he studied in the graduate lounge. Students working in labs are provided with spaces to work unlike students who do not. Individual suites within the building are newly furnished; for example the dean's suite had brand new conference furniture and office materials. However, some students hesitate to use the suite since doors and offices are not clearly marked. Faculty and staff at FIU were helpful and kind to the research team and to the students with whom we observed them interacting. They describe the environment as "like a family."

The location of the College of Engineering is clearly inconvenient for students. First, no student housing is located nearby, forcing students to drive or take public transportation from their homes. In addition, the College of Engineering is removed from the main campus, making it complicated to go between main campus classes and engineering classes. Some students describe taking the bus to the main campus to take other courses. Other students describe the difficulty in returning books to the main campus library. However, students remark that it *was* convenient that all engineering courses are offered in the same building. Nonetheless, students at FIU speak often about feeling separated from the rest of the campus. One female student explains:

> They don't pay attention to us over here, I don't think, from[the] main campus...because you go to [the] main campus and everything is...clean and shiny. It's like we don't exist. Engineers at FIU, they don't exist. They're trying to open a medical medicine school you know. I'm pretty sure they're going to spend thousands upon millions of dollars on that; and we'll still have the same sick building.

Students in the College of Engineering felt that they did not receive the same attention to maintenance and improvements as did the rest of the university.

Campus Climate and Cultural Differences

This section will include a description of student life experiences, the extent to which they received social support from faculty and other constituencies in the campus. Students at FIU engineering felt isolated from the main campus due to the distance of the engineering building from the main campus. The feeling of isolation from the rest of university was prevalent among students we interviewed. One Hispanic male engineering student comments, "We go to the main campus just to remind ourselves that there's a world out there." In spite of the detachment from the rest of the university, students describe a climate of caring and involvement from the faculty and other engineering students. In effect, the isolation, while similar to FAMU-FSU, may make students and faculty more reliant on one another:

> Everyone is very involved. They care about what the students do here. Because we're so isolated from the main campus, it's very close here. And it is a good thing. Because we can help each other out and everything. Like, for SWE [Society for Women Engineers], all of the societies are very united. We unite for a lot events and that's how we're successful.

SWE and other organizations provide opportunities for students with similar interests to come together. As a result, engineering students describe having only engineering friends. As one Black male engineering student points out, "I don't have any classes over there [the main campus]." Being isolated from the main campus has encouraged a feeling of closeness and unity among the faculty and students in the college of engineering. Despite these positive effects, the distance creates problems for students in terms of transportation, using the library, or simply socializing with other students at FIU.

Site 3. University of Florida

History

The University of Florida (UF) is Florida's "flagship" institution and the oldest public university in Florida. The University of Florida is built upon several earlier establishments including East Florida Seminary, Gainesville Academy and Florida Agricultural College, St. Petersburg Normal and Industrial School and the South Florida Military College. Located in Gainesville, the University of Florida was founded in 1853 by Gilbert Kingsbury as the East Florida Seminary, originally located in Ocala, Florida. The University of Florida was formed through the consolidation of these schools and named the University of Florida with the passing of the Henry Holland Buckman Act of 1905. Gainesville was chosen as the location. The Buckman Act created the State University System (SUS) of Florida and established the precursor to the Florida Board of Governors called the Florida Board of Control.

At the time of its establishment, UF was a male-only institution during the regular academic year (the summer session allowed women), and the first class of 102 enrolled in 1906. The College of Engineering was founded in 1910 with five faculty members and 48 students. In 1924 women of "mature age" were allowed to enroll if they had completed 60 hours at another institution and if programs of study were not offered at Florida State College for Women. UF didn't admit African American students until 1958. However, enrollment of African American students did not reach 1,000 until 1973, following a 1971 sit-in. In popular culture, UF enjoys fame through the invention of Gatorade, a sports beverage originally created for use by the Florida Gators, UF's football team. UF's total enrollment is the largest in the state and among the largest in the country with 52,084 total students, including 34,610 undergraduates. UF also has the largest College of Engineering in the state with 703 declared civil engineering majors and 446 electrical engineering majors. A Hispanic male faculty member of the college of engineering describes the greatest strengths of the program as its long tradition of good work and people, saying, "Many distinguished faculty have done work here and continue to work here. You know, UF attracts excellent students in the state." The University of Florida is a well-respected institution and one of the 62 members of the American Association of Universities, designating it as a research-intensive institution.

Student Demographics

Figure 2.7 shows that engineering students make up a larger portion of the student undergraduate population at UF in comparison with other SUS universities. UF Engineering consistently enrolls close to 13% of all undergraduate engineering students. UF Engineering also reflects the upward trend of Hispanic enrollment. Hispanic students have increased from 12% to 16% of all engineering students since 1998 at UF. While other engineering programs have seen a decline in African American enrollment, UF has held steady at 6%–7%, still the lowest Black enrollment among the SUS universities included in this research. Female enrollment has seen modest increases and decreases but remains around 20%.

One explanation for the relatively steady enrollment of UF engineering students is that they are arguably the most talented in terms of high school achievement and preparation for college-level work, thus less likely to leave the program.

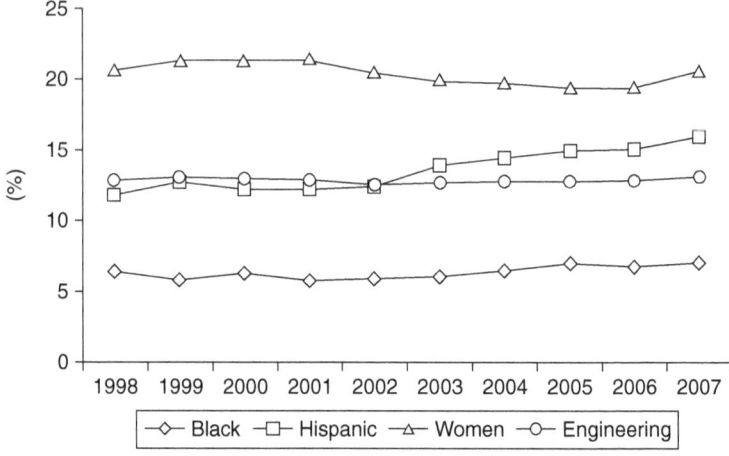

Figure 2.7 University of Florida
Source: Adapted by the authors from the Florida Board of Governors Interactive University Database.

Program Efficacy

Low Black engineering student enrollment at UF yields complex but consistent patterns of retention and persistence ratios. Black students attending UF have very low retention ratios ranging between 0.48 and 0.68. While Black students comprise between 9%–14% of freshman and sophomores attending UF between 1998 and 2007; as juniors and seniors Black enrollments encompass between 5%–7% of students enrolled. In contrast to other majors, Black UF undergraduate engineering students show high efficacy from 1998–2001, 2004, 2006, and 2007 with retention ratios around 1.00.

On the other hand, Black UF engineering student persistence ratios are lower than those for Black UF students overall. Black graduates increased from 5% of degree recipients in 1998–1999 to 8% in 2006–2007. Black engineers made up between 3% and 6% of all engineering graduates over that timeframe, resulting in low efficacy in 1998–1999, 2000–2001, and 2002–2003 to 2005–2006.

UF Hispanic engineering student efficacy levels remained constant throughout this period as increases were comparable to increases in Hispanic enrollment throughout the university. In 2001, Hispanic engineering students had high efficacy in retention along with high efficacy in persistence during the 2001–2002 school year. Persistence ratios and efficacy rates among Hispanic engineers remain low.

UF women engineers have improved retention and persistence ratios over time. Engineering enrollment and degree attainment among women has remained fairly consistent even though female enrollment and degree attainment throughout the university increased dramatically from around 50% to over 60%. Low persistence efficacy in 1998–1999 disappeared after the 2000–2001 school year for women in engineering. Women's engineering enrollment and degree attainment is generally in line with university enrollment and degree attainment.

Resources/Infrastructure

The College of Engineering at UF is housed in several different buildings spread throughout its sprawling campus. Civil, mechanical and electrical engineering are housed in Weil Hall located next to the "Swamp," or the Gators' football stadium. New buildings at UF are designed in the same red brick collegiate Gothic style as older buildings; many of the buildings are ivy-covered. However, one administrator interviewed notes that improvements need to be made to engineering buildings. Weil Hall, home to the dean's office and other major college of engineering offices, is an older building with much original woodwork. It was located on a hill and connected to other buildings. Many engineering classes at UF were taught at the New Engineering Building (NEB), about a 10–15 minute walk from Weil Hall. The NEB's center includes a study area for students, and at most times of the day and evening, these study tables are occupied. Benches line the walls and two arms of the building, and these also are usually fully occupied. A small stand sells coffee and snacks.

Campus Climate and Cultural Differences

UF engineering offers several minority initiatives, including STEP UP, a summer bridge program for minority students coming from high school into college. Due to these minority initiatives and the administrators, faculty and staff involved, minority students had a large number of positive comments about the support they received. For example, the current student president of the National Society for Black Engineers (NSBE) said this:

> I guess there's so many elements to actually making it through because you know outside the academic element. And so I guess [administrator name] would probably be the academic end. And so as far as [the] spiritual his secretary [name]; there's not a problem outside of academics that she cannot

[solve]; she'll make it happen. She'll move mountains to make sure that the students are okay outside of academics.

Gator banners and Gator paraphernalia dominate the scene throughout the campus. Many students we spoke with mentioned their pride in being a Gator, pride that came with the responsibility of living up to being a Gator. One white male engineering student recalled choosing UF because of the Gators, I've always wanted to come here since I was a little kid, growing up with the Gators watching them on TV. The football team pretty much [dominates UF's reputation]...I've always loved the Gators and it was always a dream to come here.

Two white male students discuss how they identified more with being a Gator than they did with their engineering colleagues. The first notes, "I mean the main things I get from feeling like I belong is you know, with my friends, football games, the atmosphere...That's what makes you a Gator." Similarly, the other student speaks about how he talks with his engineering cohort, "as far as work goes, yeah" but that, "but as far as you know, general things...we're all football fans." He went on to describe how being a Gator would help in industry since you can meet other Gators anywhere.

> I was walking on the beach and we just sparked up a conversation about something, about the red tide I guess and you know then during the conversation I mentioned that I was a UF student and then he goes on and on about old football games that he used to go to tailgating it is an amazing experience.

Overall, engineering students' sense of identity as a Gator was equally or more important to the students than identifying with other engineering students.

Site 4. University of South Florida

History of the University

The University of South Florida (USF) was founded in 1956 with 1,993 students enrolling for the first time in 1960. USF's Tampa campus began with five buildings on land on the very outskirts of Tampa Bay. Originally planned to serve all students south of Orlando, the University of South Florida is actually geographically north of four other SUS institutions, including Florida International University (FIU) in Miami. Touted as an urban university, the university appealed to commuter students early on though dorms we constructed between 1960 and 1965. Today, the majority of students, referred to as "resident commuters," live off campus in adjacent apartment complexes though the university is requiring all 2009 incoming first year students to live on campus in a dorm. In the last several years many additional dorms have been built on campus to accommodate larger numbers of undergraduate residents. USF is Florida's third largest university with 45,542 total students and 34,369 undergraduates. A Hispanic female administrator in the College of Engineering describes the benefits of being in a large metropolitan area, "Luckily for us, we have this whole area [Tampa] where students can engage in co-ops and that type of activity." Being situated in an area that has opportunities locally for undergraduate students is a resource that the College of Engineering is fortunate to have and takes advantage of for their students. The College of Engineering has a total of 2,613 students with 237 civil engineering majors and 242 electrical engineering majors. USF houses the second largest engineering program in the state although enrollment has remained unchanged since 1990.

Student Demographics

USF Engineering shows sharp declines in engineering enrollment compared to the rest of the university, as shown in figure 2.8. Much of this 3% decline is due to a 5% decline in female enrollment. USF is not alone in this trend as noted throughout this chapter. Also like other programs, Hispanic students have increased their enrollment in USF Engineering. African-American enrollment is consistent at around 10 percent.

When asked how representative the engineering students are with respect to gender and ethnicity, students in the focus group interviews all agree that there are more male than female engineering students. A White male student in one of the focus groups comments, "It's mostly male dominated, like probably 80%. But the females that are in the classes, they're usually the smartest in the class." A White female in a focus group agrees that the engineering program is mostly male but also points out that there is ethnic diversity, saying, "There's a large gender difference. Yeah. But as far as ethnic groups there's a lot of all ethnic groups represented." Other students interviewed confirm the large number of international students, which leads to the overall sense of ethnic diversity. Students perceive that the international students are from all over but males in one focus group comment that they are mainly from the Middle East. A White male student in a focus group summed up the general perception of student diversity at USF engineering,

> I would say here there's a lot of [students of] Middle Eastern descent, mostly males. I mean there are females in the engineering field and they do work hard at it and they're just as good as the males but then there's a lot of white males. Other than that there's some here, some there, from all the other different cultures. But I think the Middle Eastern[ers] and whites are prominent.

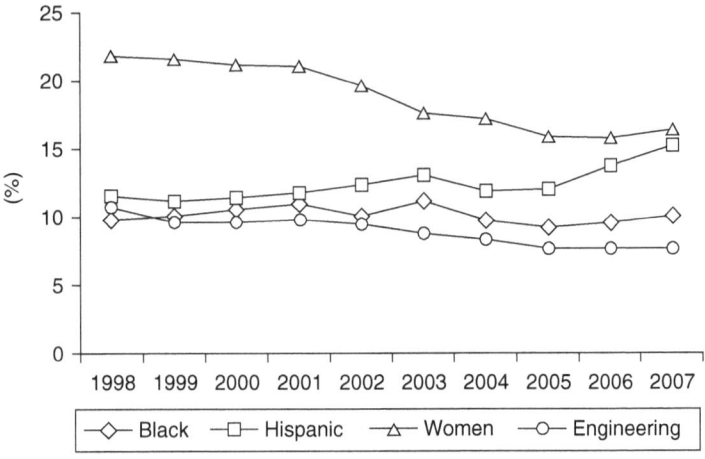

Figure 2.8 University of South Florida
Source: Adapted by the authors from the Florida Board of Governors Interactive University Database.

Program Efficacy

USF engineering has had stronger retention of Black, Hispanic, and women engineers compared to the entire university for most years from 2001 to 2006, but persistence ratios for Black engineering students were lower than the university in most years. Black engineering retention ratios have steadily increased from 0.63 in 1998 to 1.62 in 2006 showing that in those years, Black engineering students have made up an increasing percentage of upper-level students compared to lower-level students. This is an indication of a decline in Black lower-level engineering students from 13% in 1998 to only 7% in 2006. Over this period, the percentage of Black upper-level engineering students has increased from 8% to 11%. These engineering trends correspond with a small decrease in the percentage of Black lower-level students in the entire university and increase in Black upper-level students from 9% to 13%.

Persistence ratios among Black students indicate that USF engineering is low efficacy in terms of degree attainment even though the percentage of Black degree earners has improved from 5% in 1998–1999 to 10% in 2006–2007. The university has experienced similar growth from 8% to 11% over the same time period. This shows that there is a need to increase engineering enrollment among Black students.

There has been reasonable growth in the Hispanic population of USF engineering and the university as a whole. Hispanic engineering enrollment increased from 11% in 2000 to 18% in 2007, just ahead of growth throughout the university. Retention ratios held steady between 0.91 and 1.03 between 1999 and 2005, an indication that upper-level promotion is on par with lower-level enrollment. Hispanic engineering degree recipients have increased from 10% in 1998–1999 to 13% in 2004–2005 before recent declines. Persistence ratios among Hispanic engineering students are higher than among all USF Hispanic students indicating high efficacy.

USF Engineering has strong program efficacy as it relates to women, but has experienced a decline in enrollment. Women have decreased from 19% of engineering students in 1998 to only 12% in 2006. Most years between 1998 and 2006, the percentage of female upper-level engineering students was well above than the percentage of lower-level students. Throughout the entire university, women maintained a similar advantage at both levels.

Resources/Infrastructure

USF's College of Engineering is spread out in several buildings at the front of campus and the main entrance gate. The engineering buildings were built since the university's inception. Typical of mid-twentieth-century architecture, the buildings were built square and uniform; they're tan and windowless. Many of the engineering buildings connect by outdoor pathways. They are named using Roman numerals, i.e., Engineering I, Engineering II and III. The buildings surrounding the College of Engineering are older and have similar resources. In contrast, many of the buildings to the north of the engineering buildings are new and represent advances in classroom aesthetics and technology. Between the College of Engineering and the center of campus (library, Student Services, Marshall Center) there are sculptures (one of molecules which students note resembles Mickey Mouse) and newly landscaped grounds. Paths are paved between buildings but there are clear tread marks in the lawn where students have improvised more direct paths between buildings.

Campus Climate and Cultural Differences

The main engineering building has a round-about and small parking lot facing the street and facing into campus there are tall multi-storied windows. These windows look into a study area in the building. The study area, or the "Fishbowl" as students refer to it, is a social space used by engineering students to study. It is busy and full of students studying and socializing. To the left of the Fishbowl are a small Starbucks and a market where students can purchase coffee, sandwiches and snacks. There is a row of vending machines for snacks and drinks.

Students interviewed say they use the Fishbowl frequently between classes but if they need quiet to study they go to the library. Newer students use the Fishbowl as a place to meet other students. A White female student describes this:

> Actually I came into the fishbowl and sat down and the first day I saw other people studying the same things I did but I just kinda kept to myself and I [said to myself], "I don't know these people, I don't want to go intrude on them." But then you know I came here a couple other times and I'd see people by themselves and I went up to them and I said, are you studying for this? And they'd say, yeah and I'd sit with them. (White female student, USF)

Many students discussed the ease of meeting other students just by approaching them if they recognize them from class or from study areas. One Hispanic White male focus group participant remarks, "Yeah you will see a lot of students sitting everywhere so you basically go up to them and start studying with them. Most people are pretty friendly you know. Yeah you can just go [to] the Fishbowl." In another focus group, each of the three respondents report utilizing different spaces on campus. One mentions the fishbowl while the second mentions the library and the third speaks of the tutoring center. It is obvious, though, that the Fishbowl is a popular location as it was crowded every time researchers were there, even well into the evening

As mentioned previously, in addition to fieldwork conducted at four-year institutions, we also interviewed students at community colleges. While the site visits at the SUS schools were one to two weeks, the visits to the community colleges occurred over day-long visits. As a result, we will not describe the campus ecology of the community colleges but will present instead a description of community colleges in Florida and their important role as a "feeder" schools to the four university engineering programs studied in this project.

Community Colleges in Florida

Florida has formal articulation agreements between schools that traditionally feed students into the state university system, (feeder schools) and state universities. Despite formalized agreements, evidence shows that many students do not make the transition smoothly. In order to assess community college transfer student success when compared with those students who attend four-year institutions only, most research to date has used statistical measures of academic success such as grade point average (GPA) or institutional efficacy, our preferred indicator. With respect to college student performance, research suggests that community college students who transfer perform as well as their peers at four-year institutions. For example, Mattis and Sislin (2005) see little difference in bachelor's degree completion rates between community college transfer students and those students who attended four-year universities only.

Although Florida eases the transition between community college and state-supported universities by having reciprocity agreements between and among state-supported institutions, including community colleges and SUS institutions, these reciprocity agreements are not sufficient to ensure mobility for all students. Students who graduate from high school have shown they can persist through public secondary education. However, a four-year degree is considered essential for "making it" in a "flat world" (Friedman, 2005). Therefore, transitioning between high school and post-secondary school becomes critical.

In addition, Florida's community colleges and four-year programs in engineering have unique features not representative of other community colleges in the US. According to a 2005 report by the Office of Program Policy Analysis & Government Accountability (OPPAGA), Florida has fewer public baccalaureate degree-granting institutions per capita than other states (OPPAGA, 2006). This has implications for Florida students since it limits options for community college students who wish to pursue baccalaureate degrees. As we have seen, most students attending community colleges are older and less affluent. Many have families and steady jobs that make it difficult to move to enroll in a four-year program (Calcagno, Crosta, Bailey, & Jenkins, 2007). Geography has proven important in college decision making, (Micceri & Wajeeh, 1998), and students' inability to move limits their access to four-year schools.

Our study looks at community college students enrolled in courses that may lead to a STEM four-year degree. Community colleges participating in this study were purposively selected in the State of Florida because they were located in close proximity to the major four-year universities in Florida, University of Florida (UF), Florida State University (FSU), Florida State University (FSU), Florida Agricultural and Mechanical (FAMU) and University of South Florida included in our research. We focus on community colleges in a later chapter.

Conclusion

As outlined in the introduction, this research was guided by several theoretical paradigms: political economy, practice theory and an industrial/

organizational conceptualization of culture and climate. In this chapter we focused on underscoring the need for this research by contrasting Florida's student demographics with national ones and contextualizing research on college programs within these programs' cultural ecologies. We do not assume that a student will stay or leave a university solely based on perceptions of the campus aesthetics. Rather we critically examine the use of campus space through the lens of resource distribution and how resources and their use may be internalized by students. For example, large, new technologically sophisticated buildings suggest to students that departments housed within are valued by the university. Similarly, distant, unkempt buildings may make students housed there feel slighted.

Space where engineering undergraduates can gather helps students not only do engineering, but also allows them time and space to develop their identities as engineers. As we shall see throughout the volume, fostering an enduring identity as an engineer is important for student success. Similarly, departmental values may be realized or impeded by the campus ecology, for example, faculty may value face-to-face time with students but students may not be able to find their way around campus or faculty offices may not be easily accessible. We found this to be particularly true for community college students (chapter seven) who often feel lost on their new campus. Also, classroom technology may determine how faculty are able to teach: whether they can use audio and visual tools or whether they must write on the chalkboard. Chapter four, Pedagogy and Preparation, discusses how students and faculty value interactive teaching techniques but the classrooms must be able to accommodate small group work or class discussions. Students throughout our study describe their dislike of large, auditorium lectures as they feel lost or just a part of the crowd.

We saw in this chapter that students utilize social study areas such as the Atrium and the Fishbowl. These areas are important because, as we shall see in later chapters, such as chapters five and six, when academic support or resources are lacking, students are adept at relying on other students to negotiate their undergraduate programs. It is important for them to have space for study groups and to complete homework assignments. This does not lessen responsibilities universities have for supporting students; however, understanding how students interact with university resources and with each other is critical in understanding retention and institutional efficacy.

We conclude that universities cannot be compared by retention statistics alone and that even the use of institutional efficacy requires a contextualization within the cultural milieu and an understanding of how campus ecology impacts the student experience. The four programs examined here highlight dramatic differences among the universities themselves and the students who attend them. We intended this chapter to serve as an introduction to the universities studied through our research and to orient the reader to engineering programs in the State of Florida.

Chapter Three
To Stay or to Switch? Why Students Leave Engineering Programs

Will Tyson, Chrystal A.S. Smith, and Arland Nguema Ndong

Introduction

Using the perspectives of switchers, administrators, and staff, this chapter examines students' decisions to switch from engineering and enter another field. Switchers and current students, referred to as persisters, face similar problems in their pursuit of engineering degrees. Both sets of students question their ability to fit with peers and departmental values and goals. Members of the administration and staff at the four universities as well as switchers from the University of South Florida (USF) offer important perspectives on attrition in engineering programs. As discussed in upcoming chapters, persisters openly describe the challenges they are facing during interviews and focus groups. Nonetheless, they lack perspective on why *any*body would switch from engineering to another major. Persisters may depersonalize their responses, unwilling to acknowledge any obstacles preventing them from graduating. Some students are reluctant to openly discuss problems affecting their own academic performance. Current students also have limited knowledge of factors that might discourage other students from continuing in engineering. On the other hand, faculty, administrators, and staff have watched many students leave the discipline over the years and understand the reasons for departure from engineering.

This chapter relies on the institutional memory of administrators and staff as well as recent personal experiences of switchers to better understand why some students switch. Because administrators and staff manage engineering programs and are evaluated on their ability to increase program retention, they understand the reasons for student leaving. Program gatekeepers have opinions on what constitutes successful pathways to degree completion and attempt to communicate these expectations to their students.

As mentioned above, switchers are more willing than current students to discuss problems confronted in the engineering programs and relate these problems to their decisions to leave engineering. Switching is not unique to engineering. Many students change majors. More than 80% of all undergraduates change majors and some change their majors multiple times. As switching may reduce the number of upper-level math and science courses students need, changing a major may allow smoother progression to degree attainment (Micceri, 2001). Switchers have the benefit of hindsight to look back on their experiences in engineering to identify factors that led to their exit from the program. Several switchers we spoke with were awaiting graduation from USF with a degree in their new fields, fields they felt did not have the same academic stressors faced by current engineering students.

Administrators, staff, and switchers mention a number of factors in the decision to switch including disenchantment with engineering; inadequate academic preparation; insufficient academic capital; a lack of motivation to succeed in engineering; and poor social fit. We will explore these factors in this chapter.

Literature Review

In their study of undergraduate students in several science and engineering disciplines, Seymour and Hewitt (1997) determined that engineering was the most stable of the fields. That is, over half of engineering students remain in their programs compared to students in mathematics and statistics, the fields with the largest number of switchers. Seymour and Hewitt credited engineering for its relatively small number of switchers due to rigorous institutional selection and screening procedures (1997). Institutional gatekeepers, such as administrators, faculty, and staff, at the university and program levels select students for engineering programs who are perceived to have the cultural, symbolic, and economic capital necessary to persist. These gatekeepers channel students toward engineering course-taking, degree attainment, and engineering careers.

However as gatekeepers, they may also serve as barriers to attainment for students who do not meet expectations. Claude Steele's research on stereotype threat is relevant to understanding the response of professors and other guardians of institutional outcomes. Stereotype threat concerns the danger of confirming a judgment in accord with a widely held negative stereotype about the intellectual capacity of a member of one's group. In addition to their professors and administrators, students also have expectations for themselves. Students who are not successful in building peer relationships may experience social isolation and negative stereotype threat

(Steele & Aronson, 1995). As we shall see, this set of circumstances can arise for women and members of underrepresented minority groups. Steele and Aronson show that stereotype threat can result in an inefficiency of processing skills in taking on intellectually challenging problem-solving tasks, leading participants to spend more time doing fewer tasks with greater inaccuracy (1995). Clearly, this set of conditions is debilitating to individuals affected by stereotype threat including women and minorities underrepresented in engineering programs.

Chapter one introduces political economy, practice theory, and person-environment fit as the three key theoretical perspectives guiding this study. This chapter applies each theory to understand why students depart engineering. Political economy perspectives underscore the relative disadvantages that many women and minority students face in their academic preparation for engineering coursework. Practice theory examines the cultural and symbolic capital students bring with them into the engineering program setting. Cultural capital operates within the context of social relations and exchange including the knowledge of how to manage engineering as a major. If a student is the first member of his or her family to attend college, it is likely this student will not easily navigate the system necessary to complete engineering coursework. A common distinction to be made in regard to cultural capital is the comparison between book learning and street smarts. Cultural capital includes understanding the operation of institutions of higher education, local knowledge, and knowledge of program expectations and best practices that facilitate movement through engineering pathways. Social capital refers to beneficial social networks that enable students to marshal the resources they need to succeed. The third perspective examines the fit between a person and their environment, particularly the social environment that they experience through their interactions with peers as they develop a sense of belonging to their engineering program and self-identification as engineers.

Political Economy

Viewed through the prism of political economy theory, students' poor academic preparation can be understood as the consequence of an unequal education system in a capitalist society. Many students are among the brightest and best of their schools in mathematics, but attend high schools that do not have the resources to hire quality mathematic teachers to instruct high level mathematics courses such as calculus that are necessary to prepare students for college level introductory engineering courses. African American and low income students are even more likely to attend schools with fewer resources and to be taught by teachers without strong mathematic education background (Darling-Hammond, 1995; Tate, 2008).

Few Florida high school students enter college with high level mathematics and science course-taking and achievement necessary to perform well in calculus, physics, and other demanding coursework, particularly true for women and underrepresented minorities. Even among successful bachelor's degree recipients, only the few who complete high-level science and mathematics courses graduate with degrees in engineering and other science-related areas (Tyson, Borman, Lee, & Hanson, 2007). Minority persisters are challenged by the engineering curriculum: "The biggest obstacles have been just the classes...just passing the classes that is the toughest thing" (FAMU-FSU Black male persister).

Regardless of students' ability to "practice" being engineers or to "fit" in an engineering department, students must succeed academically in order to persist. Some research on engineering narrows explanations of women and minority attrition to inability to do the work and succeed in courses. But when we notice that many women and minority students switch from engineering into other rigorous STEM majors and succeed, we must acknowledge that other factors account for underrepresentation of these groups in engineering programs.

Practice Theory

The theoretical frameworks guiding this research include practice theory in addition to political economy theory. Practice theory focuses on individuals and their positions in different institutional contexts. Bourdieu (1977) contends these institutional contexts are structural constraints that influence human agency. Some suggest that Bourdieu and other practice theorists distinguish human agency from structural constraints that bear on human behavior, bounding human behavior without reducing the complexity of human activity (Postill, 2009). In other words, practice theorists argue that structures (academic, political, cultural) guide human behavior, but do not dictate behavior. Within the context of this study, practice theory posits cultural capital and symbolic capital as constituting the understanding of written and unwritten rules of behavior and social engagement expected by the academic environment and its administration.

Practice theory contends that women and minority students matriculating into an engineering program need the academic knowledge they may or may not have been afforded by their background experience, along with the knowledge of how an engineering program works in an institutional context. A competent social actor knows how to best negotiate with institutional gatekeepers to balance their own expectations with what is expected from them. Students who enter engineering with a clear sense of what faculty, staff, and administrators expect from them have the best opportunity to navigate the program to degree attainment.

Person-Environment Fit

"Fit" is a term analogous to congruence or correspondence influential across diverse areas of social science research in the last fifty years (Ostroff and Schulte, 2007). This chapter borrows from the traditions of industrial/organizational (I/O) psychology to take a closer examination of the micro-level examination of fit that focuses on the relationship of individual actors and their organizational environments. According to Ostroff and Schulte (2007):

> The basic premise of person-environment fit theory and research is that when characteristics of people and the work environment are similar, aligned or fit together, positive outcomes for individuals such as satisfaction, adjustment, commitment, performance, reduced stress, and lower turnover intentions result. (p. 1)

Of course, positive outcomes are related to persistence. Examples of successful individuals empower women and minority students to manage the unique challenges they face.

Person-side factors include personality traits and relevant knowledge, skills, and abilities but also include demographic characteristics such as race and gender. The environmental-side focuses on either the *situational-context*, such as organizational features or the *person-context*, the cumulative set of personal characteristics of individuals within the organization. This chapter examines the person-context and person-person fit is the perspective employed in this chapter. Individuals are attracted to others who have similar goals and values within the institutional context. This attraction may be a mediating factor in student decisions to persist or switch (Ostroff & Schulte, 2007).

Methodology

Analyses in this chapter are taken from semi-structured individual interviews conducted by our research team at four engineering programs: University of South Florida, Florida International University, University of Florida, and Florida Agricultural & Mechanical University-Florida State University Colleges of Engineering with a sample of switchers interviewed at USF. Participants were ensured confidentiality and data was transcribed, coded, and entered in the manner described in chapter one. This chapter examines interviews with switchers, administrators, and staff.

Switchers

To address what seemed to be underreporting of factors that lead students to switch out of engineering among persisters, we sampled all 288 students who declared engineering as their major or pre-major at USF between May 2002 and August 2006 and who were still enrolled at USF in other majors in the spring of 2008. Switchers typically exited engineering during the first or second year of college after completing lower-level prerequisites and before starting upper-level coursework. We used e-mail and recruitment advertisements in the university daily newspaper to solicit potential respondents. We conducted 17 retrospective interviews with current USF students who had switched from USF Engineering into another major. Switcher interviewees include 7 women and 10 men. Among the women are 4 Black females, 1 Hispanic female, 1 White female, and 1 Native American female. Among the men are 4 Black males, 1 Hispanic male, 1 Asian male, and 4 White males. Face-to-face interviews were conducted by members of the research team using protocols similar to interviews with current students to allow researchers to contrast responses and determine how these differences may have influenced switching.

Administrators and Staff

The research group also conducted 24 administrator interviews with engineering deans, associate and assistant deans, and 18 staff interviews with academic advisors and office assistants who interact with students on a daily basis. The primary question used to gauge their explanations for switching was "What do you think are the major obstacles to undergraduate students completing this program?" Faculty, administrators, and staff also contrasted persisters and switchers and their strategies for success when discussing other aspects of their programs.

Findings

Interest in Engineering and Academic Preparation

The overwhelming sentiment among faculty, staff, and administrators is that high powered academic preparation during high school is essential to successful engineering degree attainment. The prevailing wisdom is that students who enter college prepared for lower-level engineering courses and calculus and physics prerequisites persist and students who enter college unprepared switch out of engineering into another major. The reality is more complex. Some switchers entered college with adequate science and mathematics preparation, but switched despite their preparation. Chapter four shows that some persisters enter college without adequate preparation, but continue on to degree attainment.

Contrasts between Switchers and Persisters

Some switchers specifically credit persisters' academic preparation in high school for their academic perseverance: "I guess some [persisters] seemed more motivated, and a lot of them took calculus in high school...so they were a little bit ahead of me there...That's the only difference" (USF Asian male switcher). Another switcher recalled, "I really felt underprepared because I felt like I may not have been as strong in engineering as some of the other people that were in the program" (USF Black female switcher). Some faculty regard persisters as the best of the best:

> A lot of these kids are Bright Futures recipients. They're probably the cream of the crop in their high schools. I think they're very smart. Most of them are smarter than I am and they are very intelligent kids.... You know they're the top of the line in their science and math arena.... You're looking at the top five. Otherwise they wouldn't be here. Nobody's going to subject themselves to this kind of punishment. (FAMU-FSU faculty)

In other words, there is no point showing up at the front door of the engineering building unless you're an elite scholar, particularly in mathematics and science. Faculty see students who are prepared as noticeably different from students who are not. When asked if a lesser prepared student slipped through would be able to survive, this same professor responded:

> They wouldn't survive. They would not survive.... They can't.... See the thing about Math and Science it's a discipline, you know? And I mean they would be weeded out because that's required of the discipline. It's like music.... If you're a poor performer it's going to show. You won't be a musician very long.... Yeah you'll stick out like a sore thumb. (FAMU-FSU faculty)

This faculty member is very clear in making a distinction between those who will be successful and those who will fail, taking a decidedly deterministic stance on future outcomes for these students. Those who are prepared stay and those who are not prepared leave. Faculty, staff and administrators, as well as persisters take pride in the rigor of engineering curricula and blame switchers' lack of precollege preparation in mathematics and sciences for their attrition:

> Well, unfortunately, it's the rigor of the course work, especially for students who aren't really strong in math or science. Math and science courses are the core, the prerequisites...that you need to get into engineering, and so much of the upper level engineering course work is dependent upon mastering those subjects. If you don't do well, there's a good chance you won't persist. (UF staff)

In sum, academic preparation is from the faculty perspective the key to making the grade in engineering. Those who are well prepared in high school to take on rigorous curriculum in engineering programs at the university will likely be successful; all others will fail.

Student Expectations

Most women and minority switchers did not have enough information and did not know what to expect in an engineering program. They enrolled with limited understanding of the discipline and lacked awareness about what to expect in engineering programs. Some switchers had little idea of how their interests fit (or did not fit) with the engineering curriculum. For example, several switchers mentioned that their interest in engineering stemmed from skill in drawing and architecture, areas related to engineering, but not necessarily related to future engineering careers:

> I came to the college basically because my brother came here or whatever so I just wanted to be with him. When I got to the college, I knew I had wanted to do engineering. Well I knew I wanted to be an architect. I just didn't know which way I wanted to go. A couple of my friends I came into college with they were also doing pre-engineering and they wanted to do architecture and be architects as well, so I figured that was the way to go. (USF Black male switcher)

This student did not find the USF Engineering website helpful saying most of his information was "basically word of mouth." He later admitted that "I'm not the kind of person to get up and go try to find out about stuff I don't know about." In retrospect, he realized this hurt him, forcing him to rely on friends, also interested in architecture, who then created their own informal networks. Another switcher initially interested in architecture relied on advice from her father who had stopped short of completing the necessary credits to earn an engineering degree:

> I was maybe wanting to do architecture because I like drawing and I took... a drafting class and I really liked that but then my dad [said], "Well you should do civil engineering. That's a better degree to get." We both thought [that] he was going to get an engineering degree. But then he ended up moving down here and he was a few credits shy so he ended up just getting a business degree instead of an engineering degree. So he [said], "You should do that, they make good money, blah, blah, blah, blah, blah," whatever. (USF Native American female switcher)

Perhaps because her father switched from engineering, and indicated by her "blah, blah, blah" opinion of his advice to become an engineer, this student switched despite the promise of financial rewards.

As evinced above, the pay and job security in engineering is a major attractor for switchers. "So I probably was in engineering just because... $70,000 a year is the starting pay, so I just felt like I was just in it for the wrong reasons" (USF White male switcher). Other switchers admitted that their initial interest in engineering was not really interest at all but rather a way to enter a career in a field with many opportunities. "Basically [it was] the thought of being able to graduate with a degree that would get me a job later... It wasn't necessarily an interest. It was something that could get me somewhere" (USF White female switcher). Even switchers who were proficient in math were more motivated by high salaries in engineering than the challenges and demands of an engineering career:

> I just kind of wanted to do something with math and people told me that engineers get paid well. So back in my senior year when we were applying for schools I was kind of like I don't know what I want to do with my life and I wasn't too sure, so... I kind of threw a dart at the board and said, "yes, engineering." (USF White male switcher)

This switcher left engineering to pursue a high school education degree in order to become a mathematics teacher.

Unlike the students described above, many students who had an accurate understanding of engineering did not feel prepared because of their high school mathematics course-taking and achievement or their difficult transitions into engineering prerequisite courses. One switcher describes science fairs she attended in high school where she learned about different areas in engineering and built things from scratch in groups. After this experience, she went to her school counselor for advice:

> Basically I wanted to find out what would be the best major to go into when I go to college, because I know that sometimes people can get stuck in a major that, once they graduate they don't really know what to do with it. And so I wanted to know really what was going to make me the money when I got out of high school, because I know that was important, and so talking to my counselor, she did that for me, and then she introduced me to the programs that we had on campus. (USF Black female switcher)

Despite this knowledge, she still felt underprepared for the engineering program because she did not excel in high school mathematics courses:

> I took higher level courses in high school for math but I barely was getting by in those classes... I felt like if I had been better prepared in that area then maybe I would have stayed with engineering a little bit longer because I wouldn't have felt so uncomfortable about it. (USF Black female switcher)

While many students are lured to engineering as a major in college, they may be interested for the wrong reasons. Salary, prestige, and even skill in related areas of engineering such as mathematics do not compensate for lack of interest and lack of motivation to pursue a career in engineering.

High School Mathematics Course-taking

Switchers recount two principal narratives. Some switchers *did* take high-level mathematics courses in high school, but did not realize the importance of achievement in their ability to persist toward an engineering degree. Other switchers had the opportunity to take these courses, but declined, often due to poor counseling:

> We took science classes and when I was thinking about doing engineering I discussed it with a couple of my chemistry teachers and they [said], "Yes, it's a great field and as far as how much you are going to make when you get out" and you know that type of information....I didn't feel academically prepared out of high school for anything...Grammar, the way I was writing, my mathematical skills, all of that. I just was not prepared. I think it should be a little bit more difficult in high school because when you get to college it is like, "What? I didn't even learn this." (USF Black female switcher)

As we have already pointed out, many women and minority switchers did not have the course-taking and achievement in mathematics to prepare for an engineering curriculum, primarily because they did not know how important high school mathematics preparation would be in their pursuit of an engineering degree. Several switchers report not knowing what to expect from the university and the engineering program:

> I'm not a strong math student and these are things that my counselor knew, but I don't really feel she took the initiative to try to show me places where I can go and try to get that developed before I got to college. So when I did get to college it was very over- whelming for me. I felt like I was very underprepared in the math section. I wasn't prepared for college math. (USF Black female switcher)

High school did not equip her with the tools she needed to navigate the engineering program, nor is there a program at USF tailored to students and offering sections apart from large lecture arrangements in introductory mathematics. A strong science and mathematics curriculum in high school and effective counseling, especially in preparing core engineering courses such as calculus, is critically important for student success. But students who do not take these courses may not be prepared to start an engineering curriculum:

> Yeah, I'd say I felt very underprepared. I didn't take calculus, I didn't take...higher level math because my senior year I didn't actually, I didn't take any math my senior year.... The closest thing to help me out that I took was I'd say algebra II and geometry, but that's still...a lot more to learn after that. (USF Black male switcher)

Why did this switcher not take any mathematics courses his senior year, and then enroll in USF Engineering? Lack of appropriate counseling may be a factor, but when asked, he explained that he weighed the benefits of taking calculus, posing enrollment in calculus against the risk of poor performance in the course:

> I wanted to take [calculus] in high school to prepare me for college, but...my GPA was pretty good. It was a 3.2, something like that, in high school. So my brother had calculus when he was in high school and he ended up failing. I knew a lot of other people who tried it and failed. So I didn't take it because I did not want to put my GPA in jeopardy my senior year... (USF Black male switcher)

Calculus and other high level mathematics courses threaten the high GPAs students need in order to be accepted in a Florida four-year university as first time in college (FTIC) students. This may be a reason why some talented students do not take high level mathematics courses during their senior years. Even though he did not regret his decision to switch out of engineering, this switcher said in retrospect:

> I would have definitely taken calculus... Even further than calculus, more higher level classes to help me for college, but I didn't know what to expect back then, of course. But now that I know what to expect, I would have tried to prepare myself first before I got here. (USF Black male switcher)

Another Black switcher reports an advantage due to the cultural capital gained from his father. His father went to college and told him what to expect from college:

> Well, I took a lot of math in high school so I did take calculus in high school and [math] was my main focus, really my main focus.... I was math orientated and some sciences but definitely the math especially with calculus it definitely helped. That was probably the most beneficial.... I was lucky. My father was, you know, "you better do your math," so it was math, math, math and grades and not taking six classes of gym. And I found a lot of my friends, I mean even now, they're struggling with math and sciences. (USF Black male switcher)

He admitted that without family support he would have been struggling in his mathematics and science courses possibly impacting his switch to the business school from which he eventually earned his degree.

Some switchers who enrolled in high level mathematics and science courses during high school felt they were underprepared for college. While they may have had the technical skills, they were unable to manage college requirements. One White male switcher reported taking AP calculus and AP physics and mechanics, but being underprepared to do college-level work because his high school did not challenge him to do homework:

> I think high school really messed me up because... they're really lax on homework... They just take it if you wrote something down and they give you credit. So I never did homework... and that kind of threw me under the bus because you know I didn't know what was going on [in college]. I'd go into class and I would be so behind because I didn't do the homework but it's just something that I never did, you know. I didn't feel I was prepared to do the homework... the level of expectation was so different. (USF White male switcher)

This student believes he was underprepared because he was not well monitored to complete rigorous homework assignments in high school. In addition, he blames his lack of writing skills on his high school English course-taking, "I just felt like I've always coasted through high school especially in the [E]nglish...I took high math level classes and low [E]nglish you know and I never really learned anything in the English department." Developmental and remedial programs are expensive and have limited effectiveness; there must be a balance between institutional and individual responsibility for making the grade in college.

Transition from High School to College Science

The disconnect between high school science and college engineering is frustrating to many students. Switchers may not understand the links and applications of high school science courses to the engineering curriculum. In an ideal program, the transition to engineering coursework would involve immersing students in lab work focused on real-world engineering problems and engaging them in work with graduate students. Instead, because students do not declare themselves as engineering majors until their junior year, most are left to shift for themselves as was the case for one switcher:

> I had taken physics in high school, but it's a different caliber, a different type of physics, it's more in depth, more in detail when you get into college and it took a lot more studying and it took a lot more determination and input from me and effort (USF Black male switcher).

One switcher developed an interest in engineering based on her interest in computers and light and sound engineering from working in drama productions in high school. Understanding what was involved in pre-engineering coursework is different than having an interest in it. She admitted not researching the program requirements or spending time examining the information packet she received from USF Engineering. She said, "[it] looked like a good idea and looked like it fit.... It was what I was looking for. I wanted to work with computers. I wanted to...learn how to build them and how they worked" (USF Hispanic female switcher). Based on her experiences and interests in high school, USF Engineering seemed like a good fit; however, pre-requisites were the problem, especially mathematics courses:

> I knew I was going to have to take a lot of math and science, but I don't think until I got here that I realized what kind of math and science. So wish I would have found out what kind of math and science I was taking.

Some switchers talk about their interest and engagement in science coursework but fail to be engaged by the mathematics coursework. In this respect, interest in science does not necessarily lead to interest and ability in mathematics:

> I think that in high school I did enjoy science a lot, and I guess that was another factor with engineering is that I was good in science and I was interested in science. Math not so much. (USF Hispanic female switcher)

Although this student enjoyed chemistry and biology while in high school, she was not keen on mathematics. During the interview she remarked: "I never took physics. I wish I would have now! I wish I would have now! But I would have known...I wish I would have taken physics now." She believed that her high school prepared her; however, she lacked the rigorous coursework that would have been useful to have to fulfill engineering prerequisites. Once she matriculated into USF Engineering, she kept telling herself, "If you just get through the [pre-requisite courses], then you'll get to the good stuff that you want to learn about." Eventually this student made the decision to switch, burdened by the stress of completing physics and calculus "weed out" courses. Although she liked the elements of engineering she saw in the USF Engineering recruiting materials, the weight of coursework in mathematics was overwhelming.

Many faculty members as well as administrators support what student narratives suggest: students would be less likely to leave engineering if they received engineering experiences as first-year or sophomore students in the context of lab or summer bridge experiences. Early attrition occurs before students are fully engaged in engineering coursework, "[students are] very applied in general and they go through chemistry and math and they get bored stiff. So they drop out of engineering before they ever even get to see an engineer" (UF administrator). The foundations of engineering classes are unlikely to engage students in active learning because most courses are taught in large lecture classes. Chapter four shows that professors we spoke with agreed that students are bored with the science and mathematics foundations of engineering. "They usually get weeded out in Calc III or Physics II or something like that" (UF administrator) before they learn more about what engineers actually do. This administrator continued:

> A lot of students who are more team oriented think engineering will be too much isolation and they don't really realize that actually ability to function on teams is one of our major outcomes now that we have to insist one, ensure that students can do; so team based engineering is much more common than it ever has been but we never get a chance to tell them that so that's one of the things we're trying to put in [Fundamentals of Engineering], that if you like to work in teams, engineering is not necessarily a bad move. So that's pretty much why they leave. (UF administrator)

Switchers and persisters both complain about the science and mathematics prerequisites that delay the full transition from high school science and mathematics into engineering. Students attracted to engineering program through recruitment materials as well as from their own experience with applications and related activities may be frustrated by learning science and mathematics in non-engineering environments, "And that's what actually the students [say] after two years, 'What am I doing here? I'm studying math and science, that's not engineering.' That's one of the reasons. So they cannot relate" (FAMU-FSU administrator).

Several switchers declared engineering as their major based on the assumption their interest and proficiency in high school science would translate to engineering. Faculty and administrators admit that this transition is misleading and express regret that pre-engineering majors do not get a real engineering experience until their second or third year at the university and must survive lower-level prerequisites such as calculus and physics taught outside the engineering program that do not allow switchers to be socialized as engineering students.

The Switcher's Dilemma

Although many switchers lack the rigorous preparation in high school to guarantee successful outcomes in college, they are not alone. Many of those students who persist have similar backgrounds. Switchers have particular difficulty adjusting to first-year and sophomore year engineering coursework, but so do many persisters. Switchers face similar circumstance as their former classmates. So what factors cause switchers to finally make the decision to switch? Persisters make the adaptations necessary to succeed in engineering, but switchers do not.

Engineering Demands

Most students entering college lack maturity. While they are on their own and expected by parents and professors to be self sufficient, many students unfortunately flounder and never latch on to the college experience (Astin, 1999). Involvement in activities, especially those related to academic pursuits, but also those that engage students in ways that build bonds with the institution including participation in student government is seen by Astin and others as key to healthy development among college undergraduate students.

Administrators explain early switching as the consequence of a lack of maturity affecting students' ability to study and adapt: "They go away from their family. They find all this freedom and they cannot put up with the freedom and study at the same time" (USF administrator). Another remarked, "They don't have a right perception of how much time is required to be an Engineering student . It's just not something you do as a hobby. I mean, I tell them that all the time" (FAMU-FSU administrator). Most faculty and administrators see engineering degree attainment as a full-time job that requires a complete commitment.

Indeed, engineering faculty, staff, administrators, along with persisters take pride in the rigor of engineering curricula but admit that students generally do not understand how much work is necessary to earn an engineering degree, "It's really easy for students to get discouraged and I'm not talking about students who didn't take calculus in high school. I'm talking about the smartest National Merit kid on campus just stressing out because they're challenged" (UF staff). Interviewees consistently argued that regardless of level of preparation, all students have difficulty adjusting to engineering coursework. Many switchers and persisters are not separated by their high school mathematics course-taking, prior achievement, or ability as demonstrated by achievement in calculus and physics prerequisite courses. Many persisters in our study earn Bs and Cs and nonetheless continue toward the degree. Some persisters note that retaking courses constitutes a major obstacle to degree attainment, and pursued this as a necessary step to reach their final goals. While switchers may earn the same course grades as their counterparts who persist, switchers see coursework as a barrier instead of a challenge to overcome. This difference frustrates faculty, administrators and departmental staff:

> I was thinking about [a Black female student] who in high school, her math and science, she was at the top of her class. I mean, outscored males in her class, Black, White whatever. But now... she [says], "That's it, I don't want to do it anymore." Not that she can't do[it] anymore.... I mean she made a B in calculus and she made an A in pre-calculus and a B in calculus I or vice versa. But As and Bs. I think she made one C... I know she made a C in her physics lab and she made a C in sociology class. Other than that—As and Bs.... She said to me, "I don't want to do math and science anymore." And I'm just like.... I don't know.... Oh goodness. I'd like to understand that one, because it's not math ability. (FAMU-FSU administrator)

Beyond math ability, this administrator holds that students have to persevere through challenging coursework. Even the most talented students who enter engineering well prepared may falter, "They don't have those competencies. And skills they learned in high school don't translate to success here in college. They find that they earn a poor grade and they get afraid and they leave. I think it's a very big ego bashing to a lot of those overachiever students" (USF administrator). Another administrator from the joint program in Tallahassee noted:

> They kind of come with a false sense of security because they were pretty good in high school.... So they were able to get away with some stuff in high school...that you can't get away with in engineering when you're taking calculus and chemistry and physics and...these types of courses. (FAMU-FSU administrator)

One switcher considered himself well-prepared but ultimately switched majors because he was "sick of re-taking calculus II" and "was getting burnt out of the computer industry" (USF White male switcher). Switchers question their resolve and desire to stay in engineering as they face difficult coursework. "Everything overwhelmed me...so I [said] 'well it's only going to get harder. Is this really what I want to do and if I knew this was really what I wanted to do I would have stuck with it?'" (USF Native American female switcher). These students experience the psycho-emotional toll of burnout and feeling overwhelmed. When persistence becomes an unbearable option, switchers make the decision to leave:

> I think the biggest obstacle was basically myself. I think that's really the reason I just didn't succeed is because once I realized that this wasn't where I wanted to be, and wasn't where I wanted to go... [and] the second I realized it, I just knew that I needed to not be in the engineering building, or department anymore... But once I realized that this just wasn't my thing, I think that it was really just me deciding that, I needed to just change. (USF White female switcher)

Once the decision is made, it is agonizing for switchers to be in any way associated with the physical space much less the coursework that defines the engineering program they have left.

The Appeal of Other Majors

In addition to the push factors that compel students to reconsider their commitment to engineering, there are pull factors drawing students into other majors. Faculty, administrators, and staff understand that not all students maintain interest in engineering. For some, "their natural interests were different." An administrator gives an example of a switcher who is now "becoming an actor and he was in my class and we talked and it just turned out that this is not where he wanted to be but he wasn't going to tell anybody because his parents would be disappointed" (USF administrator).

Switching is common among all college students given their still evolving sense of self. As one administrator observes:

> Something like more than 50% of students switch their college [majors] three times, so the young people...cannot settle their mind. Maybe "my girlfriend is a business major"...all kinds of funny reasons." (FAMU-FSU administrator)

Switching may also occur because students find another major of interest thinking (as an administrator paraphrasing a student said), "Maybe I'm not good in physics, so maybe I'll do good on and take criminology, so I don't need nuclear physics" (FAMU-FSU administrator). Staff admit that a student may have simply "chosen the wrong field and they lose interest and maybe they don't get support from their parents or their parents don't know anything about it" (FIU staff). Administrators and staff understand students may lose interest, or may find other majors more compelling, but they are much less tolerant of switchers who leave engineering due to perceived lack of effort or in order to pursue another specific degree. The common narrative among administrators and staff is that switchers lack commitment, "This is engineering, so it's not going to come to you as easy as other majors" (FAMU-FSU staff). Administrators and staff deal with attrition and retention issues at the larger university level, so many openly acknowledge that engineering competes with the rest of campus for talented students. They express concern about the peer pressure students face from non-engineering students who do not face the same rigor in their course-taking,

> Well, engineering major[s] [make up] only four to six percent of [the] campus population...I'll walk you to the student union. Most likely we'll run into another non-engineering student, not many engineers because they are studying hard [and they] disappear from [the] student union. So it's a peer pressure, for one, but they cannot see this [and they think], "maybe I've become nerd or something." (FAMU-FSU administrator)

This fear that talented students may be pulled away from engineering is legitimate: "A lot of the 4.0 women leave and go into social sciences" (UF administrator). Administrators at each program report that the business school is the primary destination for switchers. "Engineers would most likely, if they drop out of here, would go to business. Very few go to other [majors]" (FAMU-FSU administrator). USF engineering to business switchers point out that it is relatively easy for engineering credits to transfer over to business, eliminating some drawbacks of switching.

Some students find that business is a better fit for their interest and preparation:

> I pulled up the curriculum for...the College of Business and I pulled up the curriculum...I was [taking in] the College of Engineering and I put them side by side...And I just felt I was better prepared in the College of Business to do what I wanted to do. (USF Black female switcher)

Many switchers regard business as a natural destination, but administrators and staff regard business as the destination for slackers who do not want to do the work necessary to be an engineer:

> There's too many students that [think], "Oh my goodness, that's just too much work. I would rather go partying with the business guys." And unfortunately that happens.... I hate to keep throwing business, but I have a business degree and as far as I'm concerned, any moron can get a business degree. (UF administrator)

One administrator describes a student who begged to take an engineering class in his first semester, but dropped it halfway through to become a business major, "So here's a guy that thinks he can take on the world but doesn't have the necessary work ethic to fulfill his own dreams or desires and would rather just slack off and take it easy" (UF administrator). At least one USF switcher believes some business switchers actually *do* see themselves as failures, "You took a math intensive degree and you had to switch to business...Maybe you see yourself as failing to complete engineering" (USF Hispanic male switcher). Administrators are also likely to disparage those who, in their view, do not have the work ethic necessary to complete an engineering degree:

> Everybody knows engineering's a tough curriculum. You're not just going to coast through engineering. People see someone coasting through business and they think that's an easy way to a good job. Now some people in business do very well but not everybody...the average starting salary in engineering is going to be higher than an average starting salary in business.... Engineering is a hard major and a lot of kids don't want to study that hard. (UF administrator)

The effect of peer pressure is strong and is felt even by persisters. One major obstacle to degree completion is "the huge workload. If you don't want to do it and...you see your other friends who are not engineers and they're not working as hard at all and you're just [thinking] 'Why am I not like one of them?'" (FAMU-FSU White male persister).

"Why Am I Not Like One of Them?"

Students struggle to meet the demands of pursuing an engineering major while also confronting the desire to be "one of them," a regular student who does not have to overcome unique obstacles to obtain an engineering degree. Engineering students work within the context of all undergraduate students in their institution, including peers outside engineering. Faculty who typically do not interact with students outside the context of engineering may not understand what is expected of students in other majors and the extent to which engineering students may wish to be typical, "one of them." Within the engineering buildings on each of the campuses students can and typically do isolate themselves within an engineering culture established by faculty, administrators, and staff. In this setting social lives outside the work of pursuing an engineering degree should not be a priority:

> If you walk around here at night you see students. This is a 24 hour building here because I've seen students bring in pillows and jugs of water during finals time and whether they're just sleeping here or waking up taking an exam studying again. It takes a lot of commitment to be in this major. (FAMU-FSU staff)

FAMU-FSU students are even more separated from campus life because the engineering program is located away from both the FAMU and FSU campuses, students may feel even more isolated from their peers in other majors compared to students at UF, USF, and FIU. The FAMU-FSU building is its own campus. "I think that once our students enter these walls they feel a disassociation from their other friends on the campuses.... They don't have that kind of student life like they have with other disciplines" (FAMU-FSU staff), which may encourage students to switch into the campus life that non-engineering students enjoy:

> And I think that first couple of semesters is like a little bit of a shock to them because Engineering is a different, a different beast if you will than maybe some of the fields that their peers are going into where they're not having to study as much. They...can participate more in the social life, the social aspects of university life. I think that's the key thing right there. (FAMU-FSU administrator)

Ultimately, USF switchers interviewed felt switching would allow them a quicker pathway to graduation and a more satisfying life, "It's only going to get harder. Why would I put myself through that when I'm supposed to be having fun in college? So I ended up switching because of that"

(USF Native American female switcher). Students come to college with an idea of what college is supposed to be, then either adjust their expectations to fit the norms of their engineering program or switch out of engineering in order to be "one of them." "It just got to the point that I wasn't enjoying myself anymore" (USF Black female switcher). Switchers also felt more comfortable in their new major. One of the positive outcomes of switching is, "...peace of mind, definitely, like it's something I knew I could do. It's something I enjoyed..." (USF White female switcher). Some switchers did wish they could have earned an engineering degree, particularly in order to earn the benefits of an engineering career and complete the task they started. Most switchers had no regrets because their switch allowed them to earn a bachelor's degree with a reasonable GPA, and, most importantly, pursue their interests, "For me, [switching] wasn't very difficult after I realized that my interest was just going and I was looking for what I was interested in" (USF Hispanic male switcher).

Fitting in with Persisters

High school preparation is necessary to survive in the classroom, but switchers point out that ability to associate with "geeks or whatever" may be necessary to fit in with other engineering students. Several switchers acknowledge that they lacked the cultural capital necessary to understand and navigate this "geek" or "nerd" culture within engineering and some attribute their decision to switch to a lack of person-person fit:

> I guess you would call them geeks or whatever but I'm not like that. I never really liked those kids you know...I would go to class with them but I would just look at them like, I didn't really want to be around people like this and that's my life. (USF White male switcher)

Black switchers see themselves as different from their mostly White and Asian peers and experience lack of fit more than students in other races. These racial differences seem to supersede gender issues for most minority females. As we shall see, female switchers seem to suggest that gender is not an issue by itself, but becomes a factor in conjunction with race/ethnicity and lack of fit with the "geek" culture in engineering.

Sociability and "Geeks"

Part of the "switcher's dilemma" discussed above results from the assumption among switchers and administrators that an active "social life" is something reserved for non-engineers. Administrators believe that student desires for a "normal" student life may hinder academic excellence and reduce motivation to persist to engineering degree attainment. Administrators acknowledge that their social expectations help students meet academic goals and several switchers linked their perception of the "typical" engineering student to persistence:

> I wouldn't say I was different, but...I'm more a social type of dude, and of course to be in Engineering, it's so much work...I like to have a social life, so...I guess that's how I differ, I like to go out and do things...I guess a lot of Engineering students in order to keep those grades up...stay home and study...so I guess that's what set me aside. I like to enjoy my life. (USF White male switcher)

The typical engineering student is described as a "quiet, non-sociable person" because engineering is "the type of program that requires you to study a great deal" (USF Black male switcher). Other switchers describe the focus that it takes in order to be a persister: "most engineering students I have seen...are really reserved, really quiet and to themselves, really focused on what is going on..." (USF Black female switcher). Switchers perceive persisters as being completed devoted to engineering in every part of their life. Switchers who did not share that commitment felt they did not fit or were different than the typical engineering student:

> I felt like I was a more outgoing personality just from the people that I saw, everybody seemed like they were the type people who leave Engineering class, and then they go to the library and they study what they just learned, and they are just all about school and all about engineering, engineering, engineering, and...that wasn't me, because I know a lot of students that just...live at the library, and that just wasn't me, but I guess that's the only way I see I was different than a typical student. (USF Black female switcher)

Administrators and staff believe that students switch out of engineering in order to find a campus life experience more similar to non-engineering students. Switchers struggle to fit in with other engineers because their social orientation typically does not match the social "prerequisites" needed for success in engineering:

> Eventually what I got hooked on once I got settled into the program was the notion of trying to live up to the reputation as an engineer, because when...I would talk to them—regular, people who were outside of engineering—they referred to us: "Oh, you are on the other side of the campus" referring to engineers, because they were so distant away from people and they thought

that section of campus was so distant from the rest of the campus. It was like there was no life over there... it was like out on the island and that is how they would refer to it. (USF Black male switcher)

Conversations like this concern administrators. Many switchers compare their experience in engineering to the experiences of friends and acquaintances in other majors around campus. In some respects, these USF switchers were socially positioned between engineering and the rest of campus before they made the decision to switch.

Switchers explain that they not only compare their experiences to those of non-engineers. They also compare their engineering experiences to their own personalities, characteristics, and social goals. "I thought I was just an introvert who just likes to work on projects but going to the College of Business I learned I'm part of that but... I like the communication part of business" (USF Black male switcher). Some switchers thought they were a good fit for engineering because of their introverted nature as they entered college. After getting to know themselves betters away from their parents and in a new college environment, several switchers realized that they were not as introverted as their engineering classmates. Another switcher compared his extracurricular leadership activities to his engineering experiences along with other things he learned about himself in college and figured out, "they didn't match. I'm a technical person, but I'm not very technical. I'm more of a social people person... and when I realized that what I was doing contradicted what my personality was and what my strengths were, then there was a conflict" (USF Black male switcher). This student felt extracurricular activities helped him realize that he is a leader and more outgoing than he thought he was before college.

Students who possess cultural capital to engage in successful interactions with engineering peers and program gatekeepers are more likely to persist in engineering. Students without the "toolkit" necessary to effectively interact with peers, faculty, advisers, and TAs find it difficult to fit in. "I didn't meet too many sociable [engineering students]. I felt out of place because I'm... really out there, I'm an extrovert and so not really feeling like I really had anybody... to identify with me... I really felt like there wasn't enough people like me as far as the sociable aspect" (USF Black female switcher). Students who fail to adapt search outside engineering for a social environment in the larger campus community that better fits their skills and personalities.

Black Switchers and Social Fit

The culture and climate in each engineering program is the result of a complex mix of race, ethnicity, gender, nationality, language and other factors or sub-factors. Jones et al. (2002) demonstrate how the climate of the learning environment may be a determining factor in college success and some educational scholars contend that African American students are dropping out of college because of lack of fit with their institution. Understanding challenges faced by minority switchers is key to understanding how race, ethnicity, and other factors shape department culture and climate and likelihood of persistence. The low number of Asian and Hispanic switcher interviewees limits our ability to discuss their perspectives in more depth. Those interviewed did feel that they fit well in USF Engineering.

Several Black switchers feel that race was an important factor in their ability to fit in the engineering program with persisters, "I didn't think I did [fit in engineering]...at all. And it's bad to say this, but because of my race" (USF Black female switcher). Another Black switcher tried to live up to the social reputation of engineers, "Yes, I felt like a minority...I'm trying this and maybe it might be or maybe it might not be...I don't know if I'm prepared or can I fit into this image or this society..." Eventually he felt like he fit after taking more classes and getting to know teachers. Other students felt that they stood out as Black engineers, for better or for worse: "I felt like I was looked at on a level, you are in engineering so you must be smart. But then I would feel like, when I don't know something I'd kind of be looked at like, 'of course she doesn't know'" (USF Black female switcher). Black switchers did not report animosity or conflict between themselves and majority White students. In fact, most Black switchers felt left out of other minority subcultures formed by international students and faculty:

> Well, I'm not a guy...and I'm not from India or even Africa or any of those countries, and it seemed like there wasn't a lot of girls in classes, there wasn't a lot of Black people in the classes...Yeah, there is a lot of Indian, there were Indian born people...I talked to one guy in there that was African, and he was from Nigeria, or Algeria, one of the two. But it just seemed there was, you know, a lot of a foreign student...and I think that's why they understood, because they can relate to the professors. (USF Black female switcher)

This switcher believes she was different because she is not a member of the primary USF Engineering subcultures, particular ones built on race and/or nationality. U.S. engineering schools tend to have a large number of foreign faculty and TA's (Seymour & Hewitt, 1997). Switchers may perceive that foreign students have the advantage of being able to relate to professors from similar cultures. It is worth noting that within the broader notion of minority status or race in America are embedded sub-factors of classification. Such sub-factors of classification are appropriated by the switcher who links her frustration with her experience in engineering to the fact that some Black faculty or students in her program are African-born. In this respect, the lack of Black American faculty impacts perceived social fit among Black students. Other switchers felt marginalized by other minority students as well:

> And the classes are very, there's a lot of guys, lot of guys, and then just with the professors there's a lot of foreign born. So I think it's a cultural thing, like in a lot of those classes... you get a lot of people from the same culture so... they just kind of stick together, and then like the other students that are not as traditionally in this major, they kind of stick together. But there's not as many of us, so you only get a couple, but I mean there were other people of other cultures that I met, I just wouldn't include them in my support system. (USF Black female switcher)

Some may contend that Black students are at fault for not including students from other backgrounds in their social and academic support systems, but this switcher points out that Black students just do what other groups do, but with less success, "I had a couple classmates, literally a couple, that I guess were my support while in there because we all didn't understand together.... It was just trying to get that 'D' or 'C' or whatever we could" (USF Black female switcher). The small number of Black students made it tough to build support groups of successful students to help pull everyone else up:

> I didn't think I fit. I guess it played kind of into the stereotype....When you look at engineering majors... one of the predominant people that you see over there are Indians. So and I guess I'm being honest, but when you go over there you see a lot of Indians and like I said they are very smart people, very smart people but when you go over there, I didn't think I quite fit into that. (USF Black male switcher)

This is an interesting take on social fit because neither switcher expresses a disadvantage specifically from being Black. The major racial-ethnic disadvantage is there are not enough Black American engineering students to form a subculture on par with Indian and African subcultures at USF Engineering. In fact, this switcher is somewhat intimidated by the stereotype that Indians are smart and felt that contributed to his lack of fit in the program. Another switcher makes a similar point:

> I think most of the people that were over there were Middle Eastern. There were a lot of them... and they were always working, they were always together working, I mean ALWAYS and that's why they were making the A's in the courses, because they really had their little community.... It

was either Caucasian or Middle Eastern, it seemed like they were predominant and mostly Middle Eastern[ers] predominate, which surprised me. (USF Black male switcher)

Some Black switchers were intimidated by the ability of students from larger groups to organize racial and ethnically segregated study groups, not easily done with similar numbers and success, also realizing that students from Middle Eastern (Asian) groups knew how to work for success in engineering:

> Later on I found out... they were just college kids like us, but just looking at it, it was like, "Oh my goodness, they are way smart and I'm like I'm not going to be here all day studying." So I was like if that what it takes, I'm not going to be able to do it and in a way that is what it took. If I would have done what they did, I would have gone through and been successful, yes. (USF Black male switcher)

Regretting his lack of success in pursuing an engineering degree, this student realized belatedly that the Asian and Middle Eastern students were "just college kids like us."

Gender as a Mediating Factor

Only a few female switchers describe gender is a major factor in their decision to switch. Black female switchers included gender in their narratives to indicate that being Black makes the demographic disadvantages faced by women and Black students that much greater, "I was a woman; there's not a lot of girls in that program, there's not. I'm African American. There's not a lot of African Americans in that program" (USF Black female switcher). One switcher believes her peers have low expectations for her because Black women are underrepresented, "I mean, you may find a female or two, but they're mostly males and then being a Black woman.... It was against me because the odds were against me because no one expected me to even major in that field" (USF Black female switcher). Another switcher agrees but claims it did not affect her personally:

> It can feel kind of isolating at times because when you look at the engineering profession, you don't see a lot of women or a lot of African Americans. Men or women. So it can feel kind of isolating and kind of like you are marching uphill or something like that, because you don't see a lot of representation. But I didn't feel like I couldn't, well at least when I started out, I didn't feel like I couldn't complete it because I wasn't like the normal... student. So I didn't feel like it was going to hinder me because I was a female minority, or even as a professional that wouldn't hinder me, but it does feel isolating. (USF Black female switcher)

While she does not believe the underrepresentation of women or Black students specifically affects her ability to complete an engineering degree, she does feel isolated, something that Black students believe several other well-represented Asian minorities do not feel.

Conclusion

Switchers have the advantage of hindsight in comparison with persisters who of course have continued with their studies. Women and minority switchers point to their inadequate academic preparation in high school as a major reason for leaving engineering. Among this group of switchers, White male switchers were more proficient in mathematics and felt they were academically prepared for the engineering curriculum. White switchers entered engineering with a more accurate view of what engineering is and what an engineering degree entails. Administrators divide persisters and switchers by their preparation but admit that lower-level prerequisites taught outside the engineering program do not give switchers a chance to fully learn how to be an engineer. Chapter four continues this discussion of how students learn to become engineers by examining faculty pedagogy and student critiques and criticisms of classroom instruction.

Administrators also judge switchers by their motivation to continue in the face of challenges. Administrators believe that all students will struggle in the classroom and switchers are merely the students who do not have the drive to survive the ego check of poor achievement and retake classes and do all the things they need to do to persist. Administrators are also concerned about the lure of other majors both in terms of academic ease and social life. For the most part, switchers are the realization of these fears. USF switchers describe their lack of motivation to continue after poor achievement or less than appealing experiences in first-year and sophomore year courses. Black switchers also describe their lack of social fit, both in terms of fitting in with the "geek" culture within USF Engineering and dealing with their underrepresentation at USF. Since we are limited to only USF switchers, we can only speculate how Black switchers fit in at FAMU-FSU with a larger population of Black engineering students or how Hispanic switchers fit in at FAMU-FSU with a smaller community of Hispanic engineering students.

In this respect, the chapter comes full circle moving from academic preparation to social and cultural capital, then explaining how Black students, particularly women, have a much more difficult time making the adaptations necessary to persist compared to other racial and ethnic minorities at USF Engineering. Black students describe a department in which principles of homophily govern the formation of study groups and relationships with minority faculty. Birds of a feather flock together, leaving the groups with fewer birds marginalized. This is a common occurrence on predominantly White college campuses (Tyson, 2002), but Black switchers suggest that other minorities are better organized and able to find strength in the margins that allow them to succeed academically and adapt socially. Chapters five and six examine the social climates and cultures of the four engineering programs.

Chapter Four
Pedagogy and Preparation: Learning to be an Engineer

Rebekah S. Heppner, Reginald S. Lee, and Hesborn O. Wao

Introduction

The curriculum and how engineering professors present it are arguably the most important components of engineering students' undergraduate experience. How well students learn, understand, and apply the curriculum largely determines their success in the academic enterprise. Indeed, the importance of improving engineering curriculum has garnered national attention.

This chapter takes into account matters related to undergraduate instruction. Using interview data from administrators, faculty and students from four engineering programs, this chapter seeks to address two major topics related to engineering education. First, we trace the development of student interest in engineering; clear differences persist between girls and boys in this process. Interest develops for boys and girls through different sets of experiences. With respect to policy implications, we know that programs targeted to women and minority students early on in their lives are avenues to developing interest in STEM generally and in engineering specifically. Second, it is abundantly clear that rigorous and relevant coursework in mathematics and science taken in high school predicts whether or not students will be on-track for STEM majors in college. We present the students' views of their preparation and of their undergraduate classroom experience and contrast them to faculty members' views. In this analysis we reveal differences between the experiences of women and under-represented minority students and their White male counterparts. These differences provide evidence of how student choices are affected by political, cultural, and class structures in society as well as by the structures of engineering programs themselves.

Interest in Engineering

Early socialization experiences for boys and young men in U.S. culture encourage them to engage in hands-on mechanical activities. Men often fall into engineering through a fascination with tools, machinery and gadgets, which they are encouraged to use more often as boys than their female counterparts (Faulkner, 2000). Childhood bonding rituals between fathers and sons in activities such as auto repair appear to create "self evident" engineering careers (Mellström, 1995) through intrinsic interest in engineering that has been found to be an important factor in persistence (Seymour & Hewitt, 1997).

Female socialization generally lacks this form of cultural capital. Research suggests that women are not as interested in tinkering with machines in the same way boys and young men tinker. Women are interested in how machines are used, not the machines themselves (Margolis & Fisher, 2002). As a result of early childhood socialization away from mechanics, women may not consider careers in technical sciences such as engineering. It should be no surprise, then, that women are more likely to become interested in engineering by participating in high school programs that target and appeal to women (Margolis & Fisher, 2002)

In contrast, African American men become involved in engineering through *both* their curiosity about "how things work" and exposure to special programs during the K-12 years (Moore, 2006). Unfortunately, minority students are less likely to attend high schools with advanced program in math and science (Untied Stated Department of Education (USDOE), 2000) Both men and women benefit from early exposure to programs that target and channel their particular forms of interest in engineering. Family influences and mentors are important to all students, but may be even more important for under-represented groups because of their less powerful positions as undergraduate students in engineering programs (Seymour & Hewitt, 1997). These personal influences complement positive images of scientists and engineers that motivate all students (Wyer, 2003), particularly females for whom seeing other women do engineering work helps break down the gender barriers (Clewell & Campbell, 2002).

Timing of Interest

Early childhood exposure to engineering is important in instilling an interest in engineering and mitigating any cultural factors that may pull women or minority students away from technical sciences. As a result, our study is concerned not only with student experiences in engineering departments but also with how students become interested in engineering.

Each student interview began with the questions: "When do you first remember being interested in engineering?" and "What about engineering interests you?" Among the 31 students who identified the age or grade level when they became interested, over half became interested in engineering during high school, rather late in the game.

Half the women in our study entered high school interested in engineering, and 25% had fathers who are practicing engineers. White students (67%) recall becoming interested in engineering before entering high school in contrast to only 21% of minority students. Students who do not have access to cultural capital because their parents are not engineers lack knowledge of the field and may cultivate little interest in engineering. Chapter three provides evidence of students who left engineering largely because they did not fully understand the field when they initially enrolled as undergraduates. In addition, students who may have some interest in engineering may not have mentors or family members to encourage that interest or to guide students through their undergraduate programs.

Childhood Interest

Consistent with Mellström (1995), Faulkner (2000), and Margolis and Fisher (2002), 41% of undergraduate men mention an interest in "how things work" or fascination with "building things" as an important motivation for entering engineering compared with only 6% of women. Our research corroborated these earlier data and we found that motivation begins at a very young age particularly for male students:

> Ever since I was a kid, I would always take apart things and, not exactly that I knew at that moment I'd be an engineer, but it's always wound up to that... If you ever did like a Lego set, my favorite thing was to not look at the instructions, so put it together based on the picture. (FIU White male student)

We found that using materials to develop spatial skills and promote visual learning can also sustain long-term interest in engineering: "I was always building like bridges and things out of Legos, popsicle sticks and things like that. But as I ascended into high school it was more computer and electrical stuff and robotics" (FAMU-FSU Black male student). Male students we spoke with traced their interest in engineering not only to their childhood interest in tinkering but also to their enjoyment of watching adults and learning from them:

> Just watching an electrician come home to work on something—maybe the refrigerator is broken—it doesn't matter. I'd always be over their shoulder

looking and basically I can memorize what they're doing....And I like reproduce whatever they can do—usually I work on my own car and I don't know—not to sound obnoxious but it just comes naturally ...sort of like it just makes sense to look at it. I can pretty much figure out how anything works just by looking at it and maybe tinkering with it just a little bit. (FIU White male student)

These examples demonstrate the impact of early socialization on future educational outcomes, but also demonstrate an early source of the gap between men and women in pursuing engineering pathways. Engaging in simple play behavior and observing engineering activities at an early age can be indicators of future interest in engineering.

Special high school programs, classes, or teachers are important influences for African American males (Moore 2006). Men and minority students were more likely to mention summer camps or other special programs providing an early introduction to math, science, and engineering. In their response to our survey, Black and Hispanic students were more likely to say that pre-college outreach training was helpful to them (70% and 57%, respectively) than were White students (46%).

Early hands-on activities are crucial to women as well. Two women describe a middle school competition building a bridge with toothpicks. Both became very involved in that competition and attribute their interest in engineering to that experience:

> In 6th grade we had a bridge building contest out of toothpicks. And I was really into it more than most of the other students. And I actually won. And they couldn't break my bridge actually..... I had a helicopter building thing out of balloons and straws and stuff. Like I really got into it and other people just kind of did it because they had to. And at the end of high school my mom was like, "you should do engineering." (UF White female student)

This hands-on experience was critical in having an important positive effect on a young woman who may otherwise have not been interested in engineering. The importance of this fascination with building things among both boys and girls has policy implications for early childhood and elementary science education. Hands-on programs for all students, particularly girls and young women, should boost interest in engineering among those who are less likely to be encouraged by parents, peers, and educators to pursue these interests.

Positive Role Models

Social science scholars consistently find the family is a key agent of socialization. More specifically, parents serve important roles as *significant others*.

We first learn to orient ourselves toward the expectations that these significant others have for us. Women in the sciences and engineering are twice as likely as men to name someone significant to them who influenced their interest (Seymour and Hewitt, 1997). Mentors and role models are much more important to young women in the sciences, engineering in particular (Frestedt, 1995). Bleeker & Jacobs (2004) found that daughters whose mothers expected them to succeed in a math career were more likely to go into a science- or math-related career.

While men were more likely than women to mention a childhood interest in engineering, women (38%) were more likely than men (22%) to mention family members or other mentors and role models who were engineers as important factors in their decision to major in and persist in engineering. A quarter of women stated their fathers were engineers, compared to only 11% of men. The importance of role models for young women is highlighted in comments made about a father as a reason for interest in engineering: "Well, my dad is an engineer. He's actually a network engineer. So I guess I wanted to follow my dad's footsteps. He's a big part of my life." But when asked later, she said she did not have any mentors:

> No. I've been on my own because I've been down here... When I say yeah my dad I think he didn't pressure me to do an Engineer at all. You know that was... That's all my decision. You know my parents backed me up 100% whatever I decide to do. But I do, I do call my dad a lot and I'm like and I'm complaining, dad this class is hard. He's like well just stick to it, you can do it you know so yeah he helps me. And he understands... (FAMU-FSU Hispanic female student)

This woman feels her parents are very supportive of her. Because of her father's cultural capital regarding engineering, however, he is able to relate to her better and she trusts his advice, as he "understands" exactly how hard her courses are. Looking outside her family for additional support another young woman recalls:

> Well, it was actually a neighbor of mine. I do have some engineers in the family... I had heard about engineering for some time, but then a neighbor of mine found out that I was going to college, and she was like, "You should be an electrical engineer like I am and then I can recruit you,"... So I'm like, "Well sure, that's uh, why not?" Then I started taking the classes that lead up to it and I was like, huh, I kind of always enjoy these classes. So I stayed with it. (UF White female student)

This student credits her neighbor, a female electrical engineer, with her decision to pursue engineering. Positive images of scientists and engineers strongly motivate students to remain in the field (Wyer, 2003) by providing examples of what young people can do with an engineering degree.

Regardless of gender and race, students from each engineering program were overwhelmingly positive about their chosen field. Over 85% of survey respondents from each program across all participating universities agreed that engineering is exciting, and engineers are respected individuals who make important contributions to society. Women and minority students, along with White male students, shared positive impressions of engineering, describing it as a field with many opportunities and work that is both appealing and challenging. Others describe the positive impact of engineers on society. Engineering is regarded as a profession requiring intelligence, analytic skills and problem solving, "That it's...a sign of an intelligent mind, I guess. You know you meet an engineer and assume that they're well read and they know what they're talking about. They solve problems." (FIU Hispanic male student)

Benefits of an Engineering Career

Chapter three shows several students who switched out of engineering were not familiar with expectations associated with an engineering degree and instead valued its earning potential. Current students mention pay and job security as part of their interest in engineering and their decision to persist:

> I'm very satisfied simply because of the opportunities that can arise after school. I mean Engineers are paid very well after you get your degree...it's a good major to have as far as job placement. It's like almost 90% guaranteed job placement because we always meet Engineers and there's always something going on. (FAMU-FSU Black female student)

Students value the engineering degree because engineering is a highly regarded profession in this society and yields future high earnings, but also demands intelligence and skill in problem solving. However, all students do not have access to the preparation in high school that required for their university coursework.

Preparation for Engineering

Minority students are less likely to attend high schools with accelerated mathematics and science courses necessary to prepare students to earn an engineering degree (USDOE, 2000). Successful completion of engineering programs also depends on student self-efficacy related to perceptions of high school preparation (York, 1994). That is, students who take and successfully complete a rigorous program of course-taking while in high school are likely to perceive themselves as competent in subsequent college-level coursework. Often students who do not complete such a program perceive themselves as competent as well. Faculty often disagree.

Student Perspectives on High School Preparation

Approximately two-thirds of students who attended U.S. high schools felt academically prepared by their high schools. Women were more likely than men to say that their high schools prepared them well academically (69% versus 50%) and that they had taken advanced placement or honors courses (69% versus 43%). This suggests that actual preparedness is related to perceived preparedness. Interviewees are primarily juniors and seniors who may be biased by their current status in the program. These students likely compare themselves to switchers who may have been less prepared or feel they were obviously prepared because they are on track to graduate.

There were only small racial differences in perceived preparation. Non-White students in our study were only slightly less likely to say they were prepared (54%) compared to White students (63%). Underserved racial/ethnic minorities are less likely to attend schools offering advanced courses, advanced placement science or second year physics or chemistry (USDOE, 2000). Non-White students may not have felt as prepared as White peers when entering the engineering program, but their success so far made them more confident in their abilities in retrospect.

Mathematics and science course-taking. Because she had enrolled in the mathematics/science track of her International Baccalaureate program in high school, a student attending FAMU-FSU picked this track "which basically focused on like programming, different computer skills, just like typing and proficiency in different Microsoft Office programs or something like that." This program ultimately generated her interest in computers and in electrical engineering specifically:

> I think it prepared me well only because of the accelerated program that I was in. So it was more project-based. We had harsh deadlines. We had to like submit papers internationally. We were tested internationally and things like that. So it was...It was more on a college level to me, my high school program....So then when I came here it was just like same thing. (FAMU-FSU Black female student)

Several students reported making choices to look outside traditional high school classes for academic preparation. Community college dual enrollment programs and the College Board's Advanced Placement program are two of the sources available for nontraditional advanced coursework while in high school: "I took lots of AP classes; took as many as I could...probably like three classes dual enrollment during my high school. I tried to challenge myself as much as possible" (UF Hispanic male student). Preparation is the result of both access to accelerated programs in high school and personal efforts to seek out advanced classes. Another student took advantage of extracurricular activities to make up for the rigid course structure of high schools in his home country, and this served as an alternative strategy:

> The high school in [my country] is a lot different from the high schools here...because we all take the same courses like we all go to the same level. It's not like here. There's no opportunity for you to grow more than your classmates because we all take the same courses. Now one of the things that I did when I was in high school was to compete a lot in...the national physical Olympics and the mathematics. So I was involved in all kinds of physics and mathematics competitions. (FIU Hispanic male student)

Students from other countries may not have the opportunities to take high level courses. A similar story was recounted by a student who grew up in rural Florida. This student became interested in engineering when she attended an engineering, mathematics, and science summer camp in the eighth grade, but she did not believe her high school prepared her for the FAMU-FSU engineering curriculum:

> I'm from the country, so all you do is basically the required classes. Like I was in gifted so...I went a little further than most students went.... I didn't have time, because I was also involved in sports and stuff. I was very active in high school, so I didn't get to go to the AP classes. But I only went up to pre-calculus. My school does in no way prepare you for Engineering or anything like that you know. It was more of an agricultural school so we didn't do that. (FAMU-FSU Black female student)

She made up for that difference by entering a 3-2 program at another university before transferring into FAMU-FSU College of Engineering explaining:

> You go to a school where like the tuition is cheaper, say you don't get accepted into a Engineering school from graduating from high school. So you go to a school where it's cheaper and you can get in and then you just do the basics there and then you do your two years to finish the engineering...

Attending another university allowed her to take courses to bridge her interest in math to the training necessary to succeed in engineering. These narratives of current engineering students provide a sharp contrast to switchers in chapter three whose interest in engineering was often vague and who frequently did not know what they would need to do on their own to prepare for an engineering program.

Faculty Perspectives on Student Preparation

Faculty, administrators, and staff members, including academic advisors, are much less positive about the adequacy of students' preparation. Faculty members were split in their views of preparation; eight said students are prepared, ten said they are not, and six said student preparation varies. One professor notes students are prepared academically but not in study habits. Administrators and staff were not asked directly about student preparation as were faculty, but when asked about obstacles to students' successful completion of the engineering degree some cited poor high school preparation.

Results varied by school, with half the positive comments from UF professors, who believed students were well prepared. None of the four administrators interviewed at UF mentioned lack of preparation as an obstacle, although one staff member, an academic coordinator, said she sees a broad range of preparation levels. UF has the highest entrance requirements of any of the programs in this study, with the result that students attending this university are likely well prepared. Faculty members universally use qualifying language such as, "*Overall* I think *in general* people are *pretty pleased* with the quality of the undergraduates that we get here." Qualifying language implies that approval of academic preparation is conditional. Several emphasize how students vary greatly in their preparedness. Approximately half the USF, FIU, and FAMU administrators in contrast included lack of preparation as an obstacle to successful program completion.

Mathematics and science preparation. A consensus prevails among faculty that mathematics and science preparation are absolutely essential to persistence in engineering. This is evinced by the following statement where a faculty member likens math and science preparation to the foundation of a house:

> In our curriculum most of our material courses require a really good mathematics background and I always tell my students that without a strong mathematical background it's really difficult to proceed. It's kind of like if you're building your house and I always tell my students, you know, if you're building your house you've got to have a strong foundation 'cause whatever you build, if you don't have a strong foundation it's gonna collapse. (FAMU-FSU faculty)

Some faculty members express frustration with the lack of student mathematics preparation, which researchers corroborate with classroom observations. These observations reveal that students were occasionally unfamiliar with rudimentary concepts. During a class session at FAMU-FSU, a professor stopped and asked a student, "Haven't you ever seen this before?" The student and the group responded, "No." The professor tried again, "I thought that you would have seen this before." The students again responded, "No." Professor pleaded: "Calculus III?" Again the students responded, "No." Such encounters seem to wear heavily on faculty:

> Yeah math is really, really awful. I mean a lot of times I'll write stuff on the board they understand the concept and I'm writing, they'll say, "How'd you get that?" Well that was just math. I mean all I did was the math part. Well I don't understand. So a lot of times it's the math that gets them, more so than the concepts in engineering, so that's one of their biggest problems. I mean I've given problems out where, the simple problems that they should know from Algebra I and I've got 40–45% of the class not getting it right. You know I'm not here to teach them math. I'm here to teach the engineering stuff. But a lot of times I go back and do some math, remedial work. (USF faculty)

Students' lack of math preparation changes the focus of the class time to remedial math and takes away from focusing on engineering concepts. Similarly, faculty members express concern that students' lack of mathematics preparation makes it difficult for them to cover difficult concepts in the classroom:

> You see some math deficiencies that are pretty substantial and I guess I see a lot of difficulty in applying theory to solving a problem, you know the problem solving seems to be a little bit on the weak side and that's something that we try to develop in the program that would sure be nice to have some more coming in so we can go a little bit further with the students. (USF faculty)

Some professors blame high schools for students' lack of preparation, "The ones graduating from local high schools... tend to be really bad in terms of math and physics" (FIU faculty). Others note that differences in student preparation are likely due to variations in high school programs, with some high schools clearly better preparing students, "I mean our students come from all over the place but I guess depending on which high school they might have gone to they might have only taken up to a certain level of math" (FAMU-FSU faculty). Concurrent research by editors and contributors of this book shows school- and district-level disparities in mathematics and science curricula throughout the state of Florida. Engineering faculty attribute students' lack of preparation to problems embedded in Florida educational policy. Some believe high school and community college instructors fail to provide adequate preparation for students who enter engineering. This holds true for those high school teachers who instruct AP courses:

> I think that [AP courses] may be misguiding a lot of students....They're not as advanced as they think that they are. Students come in with a misconception that they can opt out of certain classes and they essentially probably could; I don't think that's a good idea. For many [high school] seniors, they're not mature enough to hit the higher calculus. (UF faculty)

Another professor blames community colleges and state articulation policies that allow students to take lower-level coursework at community colleges and receive credit at four-year universities:

> And the problem is that we do have our own courses in math and physics but the problem is that most students tend to go back to community colleges to take these courses. So I see the problem in the state system of forcing universities to accept courses from community colleges. (FIU faculty)

Some engineering faculty members debate the wisdom of Florida's articulation policies allowing students to count dual enrollment coursework, Advanced Placement coursework, and community college courses toward an engineering degree. In this respect, faculty place blame on prevailing policy structures for students' lack of preparation. A discussion of pathways to engineering degree attainment through Florida's community college system including the perspectives of community college transfers, community college STEM faculty as well as university engineering faculty is explored in chapter seven.

Instruction in Engineering

This section addresses previous research on pedagogy in STEM fields, particularly with respect to successful pedagogy for women and minority students. In a seminal work, Seymour and Hewitt (1997) found that 90% of students, who switched from STEM majors, as well as 74% of those who stayed, complain about poor pedagogy in the sciences. In their study, both men and women switchers express dissatisfaction with faculty teaching. Students describe the tedium of passively observing professors working problems on the board as compared with other professors who ask them to volunteer their opinions and ideas and who provide a "sense of discovering things together" (Seymour & Hewitt, 1997, p.148). Students want illustrations, demonstrations, applications, and discussion of how lessons apply to the practice of engineering. This is particularly important for women, who may lack early socialization and hands-on experience and therefore require engagement with ideas during class (Felder, Felder, Mauney, Hamrin Jr., & Dietz, 1995). Finally, systems-thinking in engineering makes classroom engagement imperative (Chen, Lattuca, & Hamilton, 2008).

A multi-year study of active and cooperative learning in chemical engineering demonstrates improvements in both persistence and performance for participating students (Felder, Felder & Dietz, 1998). Undergraduate students view working in groups positively, with women undergraduates holding more sanguine views on group work than men (Felder, 1995). Women undergraduate engineering students are sometimes more reluctant to ask questions and participate in discussion in large groups of men, a behavior that has been attributed to their not being socialized to be assertive (Henes, Bland, Darby & McDonald, 1995). Group work can also provide social and academic support, especially important for women and underrepresented minorities as they are often left out of more informal networks (Caso, Clark, Froyd, Inam, Kenimer, Morgan, & Rinehart, 2002). Active involvement of all students in the group is critical to the effectiveness of cooperative learning pedagogy.

Active learning practices have been linked to improved student development and degree completion (Astin, 1985; Pascarella, & Terenzini, 2005). Seymour and Hewitt (1997) noted that "it did not seem to occur to most students that labs, field work and experiential classroom work could occupy a more central role in undergraduate learning [than it already did]" (p. 169). Curricular rigor is a source of pride for engineering students (Tonso, 2006), but pedagogy can be a source of frustration if students perceive faculty instruction to be "dull" or "boring" and lacking in experiences that active learning practices can provide.

How Engineering Faculty Teach

A large portion of our study was concerned with understanding practices within the classroom and how they foster student learning. Faculty and students were interviewed regarding what pedagogical techniques best fostered student learning. We will first present faculty data and then student data with implications to follow. Faculty interviewed in each of four engineering programs reported three pedagogical approaches to their teaching. Faculty members most often describe their teaching styles as interactive, emphasizing student engagement and involvement by facilitating interesting discussions. Classroom observations, however, reveal faculty rely heavily on classroom lectures and problem solving at the white board. Most faculty interviewed in engineering at FAMU-FSU, USF, and FIU report using an interactive approach while more UF faculty report using lectures. Some mention in-class group work or outside of class group assignments as a pedagogical strategy. This section examines approaches to classroom instruction and concludes with student responses and adaptations to these strategies.

Lectures. Lectures constitute standard practice across university course instruction, particularly in engineering: "I usually have a primary lecture-based thing with practice problems, in between the lectures. Because...it's an engineering kind of a class" (UF faculty). Lectures are expected in engineering and allow instructors to walk students through solving problems on the white board. This seems to be a source of frustration for many faculty who would prefer to be more interactive:

> I mean I have tried by walking all the way to the back row to just kind of get a feel and look at their eyes and see are you awake, are you with me?... But it's hard because you've got to go back and write...equations and stuff on the board. You know if it was like a lecture where there was no equations you could walk around and talk. But there's lots of writing that we have to do because of the equations and stuff that we have to put up on the board. (USF faculty)

This professor tried using PowerPoint to alleviate this problem but found it was an even worse alternative than writing at the board.

> I tried PowerPoint. Students absolutely hated it. So I threw that PowerPoint presentation out and said no more... It turns out when you do PowerPoint you're just reading it and just okay, flip, 'grrrr,' flip, 'grrrr,' flip. And the students, when it comes to that, they're like, looking at you going, "You're going way too fast. I'm not absorbing, I'm just going to tune you out." It's not a good situation. (USF faculty)

Another USF professor says he likes the "old fashioned way" of using the white board, especially in comparison to approaches such as PowerPoint. PowerPoint not only negatively affected classroom engagement and student attention, but it affected classroom attendance as well:

> I've done some of my classes in the past using PowerPoint presentations and those tend to actually harm the students if anything because well, you hand them the notes and then you come in class and you start doing it. Some of them don't show up for class. Some of them fall asleep in class because they figure they have the notes and they don't listen to what you have to say, so. (USF faculty)

Overall, faculty regard lectures as a necessary evil. However, even those who employ interactive learning and group work say they rely on lecture, believing that alternative methods are a luxury.

Interactive teaching and group work. Several faculty describe their style as "interactive," asking students questions and soliciting questions from them. Some version of "interactive" is a key instructional strategy mentioned by a majority of professors interviewed at FIU, USF, and FAMU-FSU, but only by a third of UF faculty. All USF faculty interviewed along with one UF and one FAMU-FSU faculty member emphasized student engagement using words such as "involve," "facilitate," and "entertain." An interactive class is one where "questions are encouraged, discussion is encouraged, but being a very structured, problem-solving type course, there's not a lot of debate or discussion...it's much more back and forth, question and answer" (FIU faculty). An interactive engagement approach balances the material:

> ...First of all engineering is a very dry subject. It's very dry, very much equations and math so that's not exactly something that entertains or keeps you awake so the first thing I try to do is I try to entertain. That's the first thing I have to do is entertain, keep their attention with me.... Second...they want the facts, short, sweet, to the point. I don't want to read a bunch of paragraphs in a textbook. I don't want to listen to you. Just give me the facts. So a lot of times when I give them concepts they're one line, you know, like bullets and we move on and then I do lots of problems with them. (USF faculty)

This professor believes students require simple facts. He believes in the effectiveness of a booklet filled with problems he has prepared for classroom problem solving: "They absolutely love this. That's their favorite thing in the world..." and emphasizes his use of the question "Why?":

> "Why are we doing this?" "Why are you doing this?" "Why this?" "Why that?" It's a constant why. And then I expect somebody in the class to answer me why and part of the reason I ask why is A, I want them to understand the concept rather than just crunch out numbers. Two, once you ask why attention goes up.... I do a lot of, "Hey, Jane" or "Mr. Smith, why?" "Why do you think?" "Hey Joe, can you help him out, because I don't think he understands why. Do you understand why?" [This] keeps them active and listening and thinking about what I'm asking. (USF faculty)

In this example the professor attempts to use an engaging approach, with students participating in the problem-solving process. Most engineering professors did not use interactive teaching in lieu of lectures, only using interactive strategies as an occasional alternative to lectures:

> Every now and then I'll give them problems that they solve in class. I try to ask them a lot of questions as we go. And I try to just be open to discussion in class. I don't want to be the only one talking. So I just try to have a relationship with them so that they feel comfortable to ask questions so that they don't get stuck on one point and then not hear anything else that I say. So and I think the students respond to that pretty well. I think they feel comfortable with asking questions. (FAMU-FSU faculty)

Emphasis on involvement and engagement is integral to the implementation of interactive learning and distinguishes interactive learning from lectures. Professors with an interactive style say it takes time to implement. In this respect, interactive pedagogy is a learning experience for both students and professors, "I make them go to the board and solve problems on the board. Initially they were a little intimidated by the idea but by the second or third time they loved it" (UF faculty). Gradual integration of interactive learning into the pedagogical mix gives students time to contribute to an interactive classroom discussion. "I use a mix, I use the board but I also use PowerPoint, so I try to mix it up. But it takes about three weeks to get them used to my style where they engage" (UF faculty). Patience is especially important in introductory courses:

> ...I think I'm learning and for example, the first year the freshman class I'm teaching right now, they have to be involved, because they have to call on each other to answer questions they have. To work with each other in groups they have to give presentations uh, and giving the presentations they have to call on a colleague to ask that question.... And then the lab class that I'm teaching there is a lot of interaction because you have to go around to each group to discuss.... This is for the first year students and that gets them involved in something that is in the department or in the college of engineering. (USF faculty)

After students understand that they must come to class ready to contribute, faculty can employ the interactive approach as a more structured compliment to lectures, "So we have lectures for two days out of the week and Thursday what we do is we do follow-up, students actually have to get up on the board and...they like it" (UF faculty).

Getting first-year students involved as early as possible in the engineering experience, working on and solving problems in a group context is critical. Group work reinforces efforts students should be making outside class according to a FAMU-FSU professor. This approach sets group work apart from interactive learning. Interactive learning involves the professor leading a discussion or problem solving activity; whereas group work forces students to prepare and assist their peers. This is a powerful incentive because peer relations are highly valued:

> I'll try to have a short group problem just because I don't know that they always do enough learning outside of class so I try to arrange it to force them to bring their brains to class and the only way to really have them do that is to have them work a problem without me there. So I try to do that every now and then. (FAMU-FSU faculty)

This professor uses group work to "force" students to learn because he does not expect them to do enough learning on their own. The notion that students could show up to class but not engage in active learning is what compels him to get students to "bring their brains to class." Group work allows professors to take an active role, more like an advisor, "Having students practice the material means having them work through problems during the class with me interacting and walking around the classroom while they're working on these problems" (FAMU-FSU faculty). Another FAMU-FSU professor believes that group work is "where all the learning comes from" saying:

> Sometimes I sit back and I can watch them. I mean essentially watch them learn. You know they're talking to each other, they're asking you know how did you do this, how does that work, what about this. And sometimes you know I walk around and help them out if they have questions. I find that to be the best thing.

Engineering professors have a difficult task. As discussed in chapter three and earlier in this chapter, students take science and math prerequisite courses at community colleges or in other departments around the university. Students may learn calculus and physics, but do not study them from an engineering perspective. Such learning would ease the transition into upper-level physics courses. Second, long standing norms of instructional practice underscore the use of the lecture and the white board as teaching tools. This, too, may reflect the constraints of class size even in the upper-level courses. Finally, while many would prefer being "interactive" some may lack the confidence and skill required to use this approach.

Real world applications. Group work in class often takes the form of simple problem solving exercises, but some professors introduce real world examples into the classroom especially when theory has limited utility:

> Everything I teach here at every level... is always taught with a very practical perspective in mind because it does no good to learn pure mathematical

theory. We don't do anything with that kind of stuff. It has to be something applicable in the real world. So that doesn't mean that we don't teach the theory. Of course you need to learn that but you need to learn it in a very practical context. (USF faculty)

Local communities make good locations for this applied, practical context: "I try to make it interesting...and link it to current events or local events that they can relate to...If we don't bring it very local they would never...get these concepts." (USF faculty) The university and local communities serve as a bridge between theoretical concepts and the application of these concepts in real world for engineering students:

> I often try to bring in real world things to provide context to stuff that they're learning. So we learn about this theory or idea; I can say we use this [approach] because we want to design this in the real world. So it's like an equation that they just learned and now they're using it. (FAMU-FSU faculty)

Several faculty, particularly faculty at USF, FIU, and FAMU-FSU find interactive learning, group work, and real world applications to be essential elements of their pedagogy but still rely on lecture in part, reasoning this approach is essential because of large class sizes and time constraints. Some faculty loosely integrate question and answer interactive learning into their lectures. Others work from a specific plan. Working through problems on the board and interacting with students during class engages students in learning early and often:

> I'm a strong believer in using the board. I don't do an hour and a half lecture on PowerPoint...and I think in terms of teaching engineering, having students practice the material, which means having them work through problems during the class and me interacting with them and walking around the classroom while they're working on these problems is what I do often. I often try to bring in real world things to provide content to stuff that they're learning....I would like to see 100% involvement and engagement all the time every single minute of my class. (FAMU-FSU faculty)

This professor believes that real world expectations reinforce the importance of student participation in active learning and group work exercises. Professors deem group work important, but too time-consuming to use on a regular basis. Strategies for finding time for group work include distributing lecture notes and requiring students to learn concepts without aid of lecture (Felder, 1995). More active and cooperative learning has a positive impact on students, but instructors may have to sacrifice material in order to use these alternative methods.

This sacrifice may also aid the retention of women and minority engineering student retention at the expense of the overall curriculum. Emphasis on lectures taught in the context of large numbers of students rather than on cooperative learning methods may disproportionately negatively affect women and under-represented students. On the other hand, small groups should be structured to not isolate students. Calling on students may not inspire them to discover the theory on their own. It may, in fact, add to the intimidation experienced by women and under-represented minorities who are generally are outnumbered in the classroom.

Student Responses to Faculty Pedagogy

Lectures. In short, students do not like lectures. No student interviewed discusses components of lecture-style teaching including talking from notes or using overheads and PowerPoint presentations as effective aspects of pedagogy. Ineffective lectures may result from what could be called an "engineering generation gap:"

> A lot of the professors teach the way they do because that's the way they've learned...And sometimes it's not the way other people learn. But it worked for them...The professor said, "This is how it was when I was a kid and I understood it this way. You guys are going to understand it that way." (FIU Hispanic male student)

This student is particularly critical of professors who use overhead projectors instead preferring that students be given the opportunity to work out problems in front of the class on the board "instead of having an overhead projector or already pre-made or prefab notes where they just point it out and everything because the thought process gets lost." Another student suggests additional training for professors who lecture using PowerPoint adding, "it could take a good teacher and destroy them because it is boring and when they're just reading the PowerPoint, 'blah blah, etc'. I hate that" (FAMU-FSU White male student). According to another student, students always write in evaluations that "we prefer fewer presentations like reading." She has a simple suggestion for her professors:

> Write more notes. It is easier for somebody to understand when you're writing and explaining. At least for me it is. And I feel like most professors that do that, they have better comments at the end and during the class. That is something that could be improved upon instead of just sitting there and just looking at the slides. (FIU Hispanic female student)

Students adamantly assert their strong aversion to boring PowerPoint lectures, referring to them as sleep inducing. "If you're on PowerPoint you're just like you want to fall asleep 'cause they're just pointing at stuff..." (FAMU-FSU White male student) Even though some faculty know how to present, "most of them, it's like...you go to the class just to sleep." (FIU female student) Another student complains, "A lot of us don't get much sleep. We're always studying. As soon as you turn off the lights, the heads are down, and no one's paying attention." (FIU Hispanic male student)

Interactive learning. Students prefer an interactive approach involving professors and students working through problems in "real time" and guiding them through the learning process. Students depend heavily on their interactions with their professors both inside and outside the classroom. "I think that [when] the teacher kind of interacts with the students, and don't necessarily like go like 'here's this proof, write it on the board' kind of thing, gets [a] better response from the students. You actually learn more." (UF White male student) When students interact with one another research suggests that learning is enforced in the classroom. Students generally like to work in groups, negating many drawbacks of large, anonymous lecture classrooms.

Students are clear in their preference for interactive learning. Distinguishing between "fun" classes and "tedious" classes a student remarked, "Like my materials professor makes class amazing, like really, really fun, interactive. That's good.... A lot of professors go with the PowerPoints and its a little more tedious it's harder to follow" (UF Hispanic female student). Faculty may think they are accomplishing their instructional goals, but students may not understand the concepts being taught, "They spend the whole class period showing you where the equation came from, like you don't care about that. It's not going to be on the test. Nobody's ever going to ask you where this equation came from. Like you need application" (UF Black male student). Students make a clear distinction between learning formulas and solving problems and applying these formulas, "I think it's very useless when you have a professor blabbing for [an] hour and a half and you didn't learn anything. You learned a bunch of Xs and Ys and Zs and omegas. It doesn't make any sense because you didn't see it in action" (FIU White male student).

While students across institutions draw a line between "fun" and "tedious" learning and lectures, they realize that there must be a balance between the two, "Like just lecturing sometimes can be bad, but like you know when you're asking the students too much sometimes that can be bad too. Like I don't know he's just a balance" (FAMU-FSU White male student). This balance is between the introduction of knowledge and the application of knowledge. In other words, students do not believe it is enough to learn proofs and equations. They need to know the applications of this knowledge in the real world. One student describes this balance as a "certain level of practicality" saying:

> If you're going to go into a class and say well I'm going to stress that you know how to derive this equation. When I go into real life I don't care where "F = MA" came from. I want to know...I want to know how to use the equation. That's the important thing. I don't care where it came from like. I'm not going to open a book and have a bunch of tables in front of me, I'm going to know how, need to know how to interpret. I don't need to know where the values came from. (UF Black male student)

Interactive learning opportunities coupled with real world applications constitute an approach to teaching undergraduate engineering with the most promise for involving and engaging students.

Real world application. Students express most excitement about real world application of engineering concepts. Nonetheless, only 38% of females mentioned a preference for real world examples when describing courses and professors, compared to 70% of males. This preference likely reflects the practical advantage many males have in comparison to female colleagues. Gokhale and Stier (2004) report that examples used in engineering courses assume basic knowledge of machines, making it more difficult for female students to understand relevant concepts. Only 41% of non-White student interviewees brought up "real world" examples as important in coursework, compared to 88% of White students. Again, these statistics underscore the likelihood that female and underrepresented minority students do not enter universities with the same cultural capital as do White male students.

Students see a clear link between theory and application, "Learn the theory and then you apply it. And that kind of reinforces or actually relates to the real world rather than just learning straight up formulas and how you get this and that..."(UF Asian male student). What some students like the most about their courses is that "they give kind of they give a little bit of a real world feel. The more core engineering classes where I have is all theoretical, but they kind of they have a lot of real world applications..." (UF White male student). Students appreciate "the application to the things that I see everyday....We'll talk about something and then it'll be an explanation for how a cell phone works or something like that" (UF White female student). Many students critique their classroom experience as it relates to their professional goals. They want to know what they'll need to do when they encounter situations in the real world in the practice of being an engineer. USF students describe a tough professor with a reputation for preparing students for the future saying that others say, "You'll know your stuff when you're done but you might not pass it the first time, and it's like, 'dude, I want to pass it.' I want to know it, but I want to pass it" (USF White male student). Another student describes this professor's approach:

> He's pretty hard, but the way he does it, he's gonna make you know your stuff. 'Cause a lot of the time you just take this class and it'll test you on one concept from one chapter. Then he'll take it from like three different chapters, the way you see it when you're out doing the job... Like civil engineering, you get out there and do construction. You have to support the wall where it's being built and it's just not gonna be from one chapter. You have a combination of different things and it's like what his tests are. He doesn't just want you to know one little formula and just do this and that. He wants you to know why you do it. (USF White male student)

This student understands that you would not learn what you needed to know to build a house from reading one chapter. Instead, the ability to be a civil engineer comes from applying a corpus of knowledge to the real world. Students such as this USF undergraduate engineering major relish the opportunity to understand the reasoning lying behind the application.

Student adaptations. Students experience different approaches to pedagogy in engineering courses. Courses vary greatly in the extent to which they incorporate lecture, interactive learning, group work, and real world applications, meaning that students must take their learning into their own hands because this variation may negatively impact their educational progress.

> Well a guy with a 3.0 versus a guy with a 3.5, the guy with the 3.5 is obviously a better student. But then when you look at the course of the year you might, "Oh man that guy with the 3.0 got all the hard breaks. He got all these horrible teachers over the course of his five years here." And his GPA took a major hit because of that. (UF Black male student)

Some students seem to be resigned to going at it alone with a little help from their friends, "I basically have grown to accept the fact to not expect anything from your professors and you just gotta work at it by yourself and with others" (USF White male student). Students experience frustration with professors, "because I feel like they could do more to help us learn.... I feel like some a majority of my classes are 'teach yourself' and it's hard to teach yourself engineering" (FAMU-FSU Black female student). Several students across programs talk about learning in terms of teaching themselves engineering whether it be by relying heavily on the textbook, working in groups outside of class, or just studying harder.

> The text book can really make or break the class....Our communications course has the worst textbook I've ever seen in my life and everyone's having a problem in that class because you know, like most of the time for me and a lot of my friends is we teach ourselves out of the textbook. More so than anything else and if the textbook is bad we can't follow the material. (FAMU-FSU White male student)

Improving instruction also means addressing the kinds of textbooks used in undergraduate engineering courses. In addition to an interactive pedagogical approach, the textbook is either a vehicle for learning or a point of confusion for students. Successful engineering students must develop oral and written communication skills, skills in technical report writing and in communicating ideas.

Conclusion

This chapter examines three components of the process of learning to be an engineer: interest, preparation, and classroom instruction. Gender plays an important role in initial interest in being an engineer. Young men are likely to become interested in engineering through early socialization; the activities that they are encouraged to pursue allow them to discover they enjoy learning how things work. Women we interviewed, on the other hand, only became aware of engineering careers through special school programs or because they had a role model or family member in the field. As we will evince in chapter six, mentors and role models in the sciences are also less available to members of under-represented minority groups. This difference in early experiences is an example of what Bourdieu called "the role of social origins," in the creation and maintenance of inequality (Robbins, 2005). College outreach programs and first or second year introductory courses were shown by our survey to be more helpful for non-White students. These programs and courses can be seen as a way to address the gap caused by less access to cultural capital before college.

Engineering students in our study are much more positive about their high school preparation than faculty and administrators. This confidence, or self-efficacy, is likely important to successful completion of engineering degrees; however, if it is combined with under-preparation, serious challenges confront both students and their professors. This combination of over-confidence and poor preparation has been linked to failure to persist by under-represented minorities in science (Seymour & Hewitt, 1997). Our research similarly indicates this contradiction in perceptions of preparedness. Several faculty and administrators said that students are in fact under prepared; most programs allow students to take extra coursework to address the problem. However, many students assert that they are well prepared by virtue of enrolling in Advanced Placement, honors courses, gifted or International Baccalaureate programs. In fact, research demonstrate that under-represented minorities overall are less likely to attend schools that offer such programs (USDOE, 2000). Without access to accelerated programs, students will struggle to compete as undergraduates in engineering. Furthermore, faculty question whether or not AP courses prepare students as much as they should.

Women were more likely to say they are better prepared than the men whom we interviewed, a finding consistent with the NCES study of persistence that found women more likely to complete engineering programs (USDOE, 2000). Women who do *not* feel they are well prepared may not attempt to major in engineering, while males, perhaps, are more willing to risk failure. Again, this harkens back to women's early socialization, lack of early exposure devoid of engineering and STEM fields, and lack of exposure to engineering before college. Non-white students were less likely to say they were well prepared and ready to take on a major in engineering. This difference is symptomatic of their relative power positions in the larger society.

Students expressed strong preferences for interactive teaching and real world examples and marked distaste for lectures. We agree with Seymour and Hewitt (1997) that lack of a demand by students for truly "active learning" is most likely because they have had little, if any, experience with it. Students are more critical than professors of the status quo, but still seem to accept that engineering classes have to be configured a certain way. When professors show enthusiasm, students are ready to overlook their mistakes, as long as professors are willing to show them in "real time" how problems can be solved. This seems a fairly low standard for pedagogy, considering the preponderance of research on effective teaching (Astin, 1985; Felder et al., 1998; Pascarella & Terenzini, 2005).

We find only slight differences in pedagogic preference between the majority, White male, students and those of under-represented groups. The preference for "real world" examples is expressed more often by White students than non-White, and more often by males than females. Women may be less comfortable with real world examples because examples chosen by predominantly male engineering faculty are likely to be from realms they have not experienced, such as mechanics or construction. In the case of minority students, those we interviewed are interested in real world examples, but this is less important to them than other issues regarding coursework, such as the challenges they face in gaining access to required courses. These challenges are further evidence of the impact of their relatively lower levels of power and cultural capital.

In summary, our findings show that women and under-represented minority students have relatively less power- cultural and social capital- when entering engineering fields, which continue to be dominated by White men. This relative positioning is evinced in the classroom where women and minority students may be underprepared or unable to contextualize what they are learning in the real world. The use of cooperative and active learning methods is an approach that engages women and minorities in classroom instruction, and is one way to overcome this lack and improve graduation rates for these students. There is also a clear need for proactive approaches, such as special high school programs, to encourage both of these groups to pursue the sciences. For those students who do not have family members in the engineering profession, the limited availability of female and minority role models reduces the number of students from these groups who pursue the field. Women who are not well-prepared by their high school education are less likely to take risks and pursue STEM degrees including degrees in engineering. Unfortunately, under-represented minorities are less likely to *be* well-prepared, because they may lack access to accelerated programs in high school. Improving access and encouraging participation among all students in those programs should be a priority for the math and science education reform agenda.

Chapter Five
Program Climate: Engineering Social and Academic Fit

Hesborn O. Wao and Reginald S. Lee

Introduction

Departmental *culture* refers to the fundamental ideologies, assumptions, and values espoused by members of an organization. Department *climate* is student perceptions of "what it is like" to be in the department in terms of practices, policies, procedures, routines, and rewards. Additional aspects of climate include feeling a sense of belonging, identifying with other members, and feeling comfortable in one's environment. Climate originates from a psychological framework, specifically industrial/organizational (I/O) psychology, and culture has anthropological roots as described in chapter one (Glick, 1985). Constructs related to climate refer to experiential descriptions of *what* happens in the department while culture examines *why* this climate exists (Ostroff, Kinicki, & Tamkins, 2003).

A mixed-methods approach is used to examine the climate of engineering departments as this approach uncovers complex issues such as racial/ethnic and gender bias in higher education. Quantitative analyses yield three factors that predict intent to leave: institutional support, personal agency and peer support, and perception of fit. These analyses show no statistically significant differences for race or gender, but qualitative analyses yield specific themes for women and minorities. Overall, students report that women and under-represented minorities are not treated differently; however, their narratives suggest they indeed do experience departmental climate differently than their white male counterparts. Thus, racism and sexism is subtle such that students who experience it are often unable to articulate it. To fit into the department academically and to fit with peers socially, female and minority students must develop strategies to acquire the cultural capital necessary to succeed in engineering programs. Women and underrepresented minorities frequently adapt to fit into the engineering department by creating spaces for themselves in student organizations.

The "Climate for" Approach

An engineering department may be viewed as an organization with members including students, faculty, administrators, and staff. Members interact with each other on a daily basis, creating practices and routines unique to the particular department and its programs. This chapter adopts the "climate-for" approach to hypothesize that engineering programs can build a climate for student program efficacy and student graduation. This approach is consistent with previous research showing that active learning, collaboration, participation, and mutual respect are aspects of climate related to student program efficacy. Schneider (1990) proposes that climate is a construct with a particular strategic focus that reflects the organization's goals, thus acting as a specific outcome such as "climate *for* service." The "climate-for" approach has gained a great deal of support in recent years as researchers have studied issues such as climate for safety (Neal et al. 2000, Neal & Griffin, 2006), service (Salvaggio et al., 2007, Schneider et al., 1998), and justice (Naumann & Bennett, 2000; Yang et al., 2007). This chapter examines factors that constitute a climate for program efficacy as it pertains to retention of women and minorities in the engineering programs at selected universities in the state of Florida.

Climate for Program Efficacy

Tinto (1975, 1993) holds that students' levels of academic and social integration into an academic department influence levels of commitment to their goals and institutions, which in turn determine their likelihood of persisting to degree completion. Astin (1984, 1999) builds on Tinto's theory by stating that students' involvement increases their satisfaction with their college experiences (e.g., interactions with faculty members and participation in student organizations), which in turn, impacts the likelihood of persisting to graduation. Critics contend that Tinto's focus on academic and social involvement does not adequately explain persistence among female and minority students, suggesting that student success is contingent on the degree to which they can separate from their traditional cultural traditions, values and customs and adopt those of the primarily White culture of their college campuses. In this respect, the institutions themselves must be responsible for integrating students into their campuses (Attinasi, 1989; Kraemer, 1997; Rendón, Jalomo, & Nora, 2000; Tierney, 1992).

Practice theory examines individuals' social engagement within their settings (Bourdieu, 1979), in our case, the settings female and minority students experience in engineering departments. Our understanding of *habitus* as a theoretical construct is that students cannot be understood apart from their social and physical setting but are in fact embedded within it. We examine student report of the ways universities and departments provide support to women and minority students and how they create a climate conducive to these underrepresented groups to succeed. Departments may be supportive of female and minority student organizations where students learn important skills to succeed in engineering, thus providing students agency in the involvement process (Astin, 1984) or integration process (Tinto, 1993) rather than reducing them to being merely passive members of the system. Departments and programs that focus on collaboration rather than competition, are collegial rather than bureaucratic, and are student-centered rather than institution-centered, tend to be associated with increased success for all students, particularly female and minority students (Nixon, Meikle, & Borman, 2007; Smith, Gerbick, & Figueroa, 1997; Tinto, 1993).

Hall and Sandler (1982) originally coined the term "chilly climate" to describe faculty members' often unconscious behaviors contributing to classroom environments that disadvantage women. These include behaviors such as calling on male students more often than female students, paying more attention when men speak, and focusing more on a woman's appearance than on her accomplishments. Later, Hall and Sandler (1984) expanded this idea beyond the classroom to the "chilly campus climate." Prior research suggests that such behaviors and the environments they create often go unnoticed because they reflect socially accepted patterns of communication and the long-held belief that men are more capable of working in the fields of the so-called hard sciences (Brady & Eisler, 1999; Sandler, Silverberg, & Hall, 1996). Seymour and Hewitt (1997) build upon this idea, suggesting that the "chilly climate" has led to increased self-doubt in women resulting in their attrition from engineering fields.

The primary research question in this chapter is: "Which specific climate differences between programs and schools are associated with these programs' success in graduating females and minority students?" Much of the research on organizational climate throughout the 1960s and early 1970s focused on relationships between climate and organizational outcomes such as performance, satisfaction, stress, commitment, helping behavior, turnover intentions, absenteeism, and involvement (Ostroff, Kinicki, & Tamkins, 2003). In the 1980s, controversies arose regarding the objective versus the perceptual nature of climate, the appropriate level of analysis for addressing climate, and the aggregation of climate perceptions

(Ostroff, Kinicki, & Tamkins, 2003). Following these debates, it is widely accepted that the measurement of climate must begin at the individual level (referred to as psychological climate), but can be meaningfully aggregated to represent organizational climate when there is consensus among individuals' perceptions of climate (James, 1982; James, Demarie, & Wolf, 1993; Ostroff, Kinicki, & Tamkins, 2003).

Measuring Program "Climate": The Use of r_{wg} Statistic

A quantitative climate construct was introduced in the 1960s based mainly on the work of Kurt Lewin who studied the climate created by different leadership styles and the consequences these different climates had on the behaviors and attitudes of group members. Lewin (1951) argued that climate is perceived by individuals, but can be measured and studied at a group level. Departmental climate is present if there is an acceptable level of agreement among students across all dimensions of climate (Kozlowski & Klein, 2000; Lindell & Brandt, 2000). Nonetheless, groups within departments may develop unique climates, that is, the *content* of climate can vary across groups within the organization (Lewin, 1951; Schneider & Bowen, 1985).

There is currently no consensus among researchers on a single suitable statistical index for assessing climate throughout an organization. This study uses r_{wg}, a measure of inter-rater agreement (or variability) developed by James, Damaree, and Wolf (1984, 1993), to determine if student ratings are sufficiently homogeneous to justify aggregation (e.g., to compute average values of the climate measures). This index compares the observed within-group variances to a theoretical reference distribution, that is, $r_{wg} = 1-(S_x^2/\sigma_e^2)$, where S_x^2 is the variance of the observed ratings, and σ_e^2 is the expected variance when there is no agreement among the raters. An r_{wg} value equal to zero indicate no agreement among raters, a value of 1.0 indicate perfect agreement, and values greater than 0.70 are considered sufficiently high to justify aggregation of individual responses to group-level measures. In this study, high values of r_{wg} suggest existence of organizational climate (Lindell & Brandt, 2000).

Method

Design

This study employs a mixed methods approach to understand aspects of climate in engineering departments or programs that enhance student program efficacy and completion of engineering degrees for women and underrepresented minorities. The quantitative component of our investigation involved surveys with students at four selected Florida engineering programs and the qualitative component involved individual interviews and focus groups with students at the same institutions. Within this partially mixed concurrent equal status design both components were undertaken concurrently and weighted equally in addressing issues related to department climate, mixing only at the data interpretation stage (Leech & Onwuegbuzie, 2009). The words and narratives obtained from thematic analysis of the qualitative data provided a strong complement to the quantitative findings.

Participants

Quantitative data include a survey administered to 881 current engineering students at the five institutions (25% female), the majority (86%) of whom were either in their junior or senior years in college. Participants in the student survey were mostly White (43%), followed by Latino/Hispanic (27%), Black (18%), Asian/Pacific Islander (7%), and Other (5%). The qualitative data included 44 student interviews (36% female) and eight student focus groups comprising 29 participants (21% female). The survey, interview, and focus group samples were representative of the fall 2007 student enrollment with respect to gender and racial/ethnic composition.[1] Chapter two contextualizes the survey, interview, and focus group data by detailing the basic features of the institutions including campus ecologies and resources.

Measures

The 73-item student survey contained nine subscales that represent nine dimensions of climate used to evaluate intent to leave as a measure of program efficacy (O'Reilly, Chatman, & Caldwell, 1991; Ostroff, Kinicki, & Tamkins, 2003). Each scale was developed using five items anchored on a 5-point Likert-type scale, with measures of agreement ranging from 1 = strongly disagree to 5 = strongly agree. Other Likert ranges with different definitions are indicated in the descriptions below.

Support and Facilitation Scales

Involvement. This includes student perceptions of faculty involvement in the academic life of the department using agreement with a set of items capturing faculty availability and help to students, responsibility for student success, and enthusiasm about teaching. Examples of items included "faculty and staff help students achieve professional goals" and "faculty members are enthusiastic about teaching." The involvement scale had internal consistency of 0.70 in our sample as measured by Cronbach's alpha (α), a numerical coefficient of reliability in which higher values indicate reliable scores and vice versa.

Faculty support. This was measured with four items for which students were asked to indicate their agreement with a set of items capturing the types of assistance provided by faculty and staff to help students' master knowledge in their discipline and develop creative capacities. An example of an item in this scale was "faculty and staff provide students with strong academic and professional role models." ($\alpha = 0.76$)

Institutional support. This measures perceptions of the support and services provided by institutions to help students succeed in school such as pre-college outreach or training or tutoring support. Students responded to eight items by indicating how helpful they found the listed service ranging from 1 = very unhelpful to 5 = very helpful. In addition, students indicated if they "did not participate in the service but it was available" or their "institution did not offer the service." ($\alpha = 0.72$).

Cooperation and Harmony Scales

Helpfulness. This included six items used to assess the extent to which students perceived members of the department as helpful by indicating their level of agreement with items such as "people generally care about students' wellbeing," "the interpersonal atmosphere is cold," and "faculty and staff make students feel inferior." ($\alpha = 0.71$)

Diversity. Twelve items included in this subscale capture the extent to which students perceived that members of their department embrace diversity. The first set of nine items asked respondents to indicate their agreement to statements regarding what happens in the department. The remaining three items prompted students with the question, "Since coming to the department, how often have you done the following?" to which they responded by indicating the frequency with which they carried out activities such as "Working in small, ethnically diverse groups with other students in the department" or "Socializing with someone of another race or ethnic group" on a scale of 1 = never to 5 = very often. ($\alpha = 0.74$)

Integration Scales

Integration. Within the department this is measured by agreement with six items such as "students often work together on team projects," "students share strategies for success with each other," and "students often learn from each other." ($\alpha = 0.65$)

Fit. This includes two items that measure students perceived fit in their department using agreement with "I feel like I fit in well" and "I sometimes feel out of place." ($\alpha = 0.61$)

Engagement. This measures the extent to which students perceived they were engaged in their academic work using agreement with three items: "Students are encouraged to develop critical, evaluative, and analytical qualities," "Students are highly engaged in coursework," and "There is an emphasis on developing vocational and occupational competence." ($\alpha = 0.61$)

Importance. This includes student perceptions of the importance of their field as assessed by indicating their level of agreement using seven items. Examples included "Students have to study very hard to succeed," "Students are well prepared to obtain very good jobs when they graduate," and "Individuals getting a degree in my major are respected by most people." ($\alpha = 0.73$)

Intent to Leave: A Measure of Program Efficacy

To measure intent to leave, students were asked to respond to the statement, "Given an opportunity to enroll in the same degree program at a different but equally ranked university, I would..." by indicating whether they would (a) *definitely* maintain enrollment at their university, (b) *probably* maintain enrollment at their university, (c) don't know—no opinion, (d) *probably* enroll at the alternative university, or (e) definitely enroll at the alternative university. This item was then dichotomized into "not leave" (i.e., definitely or probably maintain enrollment at their university) and "leave" (i.e., probably or definitely enroll elsewhere) for use in the logistic regression analysis.

Analyses

Quantitative analyses first compared student responses on individual items which were then aggregated into the nine theorized measures of climate by institutions, gender, and race/ethnicity. To do this, we computed *Cronbach's alpha* and *item-total correlation*, the relationship between an individual item and a composite factor whereby high correlation values signified items to be retained as they seem to measure what the rest of the items in the subscale measured. Next, we computed the r_{wg}, measure of interrater agreement to determine if student-level data were sufficiently homogeneous to justify aggregation. To determine the extent our data supported the theorized nine measures, we conducted confirmatory factor analysis (CFA) [2] using the principal axis factor method. Because the factors obtained from the theory-driven CFA were not predictive of "Intent to Leave," we then conducted a data-driven exploratory factor analysis (EFA) which yielded three factors predictive of *Intent to Leave*, a proxy for program efficacy.

Qualitative analyses were conducted using the constant-comparative method described in chapter one. Results from quantitative and qualitative data analyses were integrated in a coherent set to illuminate the understanding of program climate associated with successful program efficacy and graduation with engineering degrees. Because our correlational design was not aimed at establishing causality, we relied on the literature to identify factors that have been found to be related to student program efficacy and to discuss the extent to which our data supported such relationships.

Results

We present the findings in two parts. The first part, largely based on survey data, includes means, standard deviations, correlations among climate measures; the extent of agreement on the measures by institutions and by programs; and differences in student perceptions by institution, gender, and race/ethnicity. Next, insights from qualitative data helped to explain the three factors predictive of *Intent to Leave*, our proxy for program efficacy in this study.

Over 60% of the 36 bivariate positive correlations among theorized climate measures were at least moderate ($.40 < r < .73$), the highest correlation being between *Involvement* and *Faculty Support* suggesting that students perceive faculty as being more supportive as faculty involvement in students' academic lives increased (figure 5.1).

Having established that the measures were correlated, we computed r_{wg} to determine if student-level data were sufficiently homogeneous to justify aggregation. Figure 5.2 shows agreement among students at the institution level and at the department/program level in bold.

Except for *Fit*, where the r_{wg} values at each institution ranged from 0.63 to 0.67, the r_{wg} values for the remaining climate measures were at least 0.82, indicating sufficient agreement among students at the remaining institutions to justify aggregation. Although perfect agreement on *Institutional Support* was noted at the four programs, an interesting finding was that there was no agreement in two departments (electrical and civil) at FIU and four departments (civil, mechanical, other, and electrical) at UF. The two low rwg values of *Intent to Leave* (0.22 at FAMU/FSU

Figure 5.1 Means, Standard Deviations, and Correlation among Theorized Climatic Measures (N=867)

Measure		F1	F2	F3	F4	F5	F6	F7	F8	F9
F1	Involvement	–								
F2	Faculty Support	0.729	–							
F3	Institutional Support	0.176	0.172	–						
F4	Helpfulness	0.557	0.596	0.117	–					
F5	Diversity	0.422	0.432	0.113	0.572	–				
F6	Integration	0.284	0.292	0.086	0.521	0.458	–			
F7	Fit	0.259	0.310	0.086	0.486	0.400	0.484	–		
F8	Engagement	0.491	0.529	0.103	0.531	0.503	0.438	0.321	–	
F9	Importance	0.483	0.497	0.208	0.472	0.486	0.402	0.334	0.618	–
Mean		3.562	3.552	2.812	3.617	3.672	3.99	3.663	3.862	4.02
Standard Deviation		0.619	0.695	0.999	0.624	0.527	0.593	0.865	0.64	0.566

Note: Departments with fewer than five respondents were omitted

and 0.20 at USF) do not limit our ability to aggregate the data (Lindell & Brandt, 2000).

Next we determined that the nine theorized climate measures were intercorrelated and that student data could be aggregated. We then conducted a confirmatory factor analysis which yielded nine data-driven factors. We correlated these with the theorized measures to determine if the two sets of measures were related. Except for three factors from the data-driven results that did not distinctively capture specific theorized climate measures, over half (42) of the 81 zero-order correlations were at least moderate ($.40 < r < 1.0$) indicating that the results of the two factor analyses were comparable. For example, Factor 1 of the data-driven result included *Involvement* and *Faculty Support* of the theorized measures.

Differences by Institutions and by Gender

Preliminary analysis (not shown) revealed that there were no statistically significant differences in the theorized climate scales by institution (based on one-way analysis of variance) or by gender (based on independent samples t-tests). *Importance* had the highest mean in each of the four institutions (mean of approximately 4.00) indicating that most students surveyed viewed the engineering major as being very important. Although they agreed that they had to "study very hard to succeed," they also agreed that "the degree they are working on is in an exciting field," "individuals

Figure 5.2 Means, Standard Deviations, and Agreement on Climate Measures among Students by Departments and Programs

Department / Program	η	Theorized Climate Measures									Intent
		F1	F2	F3	F4	F5	F6	F7	F8	F9	
FAMU/FSU	**237**	**0.88**	**0.86**	**1.00**	**0.87**	**0.90**	**0.86**	**0.64**	**0.83**	**0.91**	**0.22**
1. Civil	48	0.91	0.90	0.21	0.88	0.91	0.87	0.76	0.83	0.91	0.38
2. Mechanical	67	0.88	0.85	1.00	0.85	0.90	0.89	0.72	0.84	0.90	0.26
3. Electrical	44	0.85	0.84	1.00	0.88	0.91	0.85	0.60	0.79	0.91	0.18
4. Other	22	0.88	0.90	1.00	0.88	0.90	0.87	0.43	0.88	0.91	0.16
5. Computer	17	0.85	0.87	1.00	0.87	0.90	0.86	0.72	0.86	0.91	0.13
6. Chemical	31	0.84	0.77	1.00	0.87	0.92	0.89	0.59	0.85	0.94	0.00
7. Computer / Other	6	0.88	0.88	1.00	0.80	0.75	0.37	0.60	0.74	0.90	0.00
FIU	**183**	**0.85**	**0.82**	**1.00**	**0.85**	**0.92**	**0.86**	**0.63**	**0.82**	**0.90**	**0.20**
8. Computer	26	0.83	0.83	0.28	0.83	0.90	0.90	0.58	0.84	0.91	0.49
9. Other	43	0.88	0.86	1.00	0.86	0.90	0.84	0.56	0.81	0.89	0.25
10. Electrical	40	0.88	0.82	0.00	0.85	0.93	0.88	0.65	0.86	0.91	0.16
11. Mechanical	35	0.85	0.83	1.00	0.88	0.92	0.79	0.72	0.76	0.87	0.14
12. Civil	33	0.82	0.75	0.00	0.78	0.90	0.85	0.59	0.78	0.89	0.04

UF		F1	F2	F3	F4	F5	F6	F7	F8	F9
	231	**0.88**	**0.87**	**1.00**	**0.88**	**0.91**	**0.89**	**0.63**	**0.86**	**0.92**	**0.65**
13. Environmental	8	0.72	0.51	1.00	0.64	0.61	0.74	0.30	0.56	0.43	0.88
14. Civil	124	0.89	0.90	0.00	0.90	0.93	0.90	0.67	0.86	0.93	0.77
15. Chemical	8	0.87	0.79	1.00	0.89	0.95	0.91	0.43	0.84	0.92	0.75
16. Mechanical	23	0.86	0.86	0.00	0.89	0.90	0.90	0.47	0.86	0.94	0.73
17. Computer / Other	9	0.90	0.82	1.00	0.84	0.83	0.81	0.58	0.95	0.88	0.50
18. Other	31	0.90	0.85	0.00	0.88	0.91	0.88	0.63	0.86	0.91	0.50
19. Computer	7	0.86	0.80	1.00	0.62	0.94	0.77	0.43	0.89	0.94	0.25
20. Electrical	11	0.83	0.80	0.00	0.84	0.92	0.93	0.63	0.87	0.95	0.19
21. Undecided	5	0.82	0.77	0.24	0.90	0.95	0.93	0.73	0.89	0.95	0.00
USF	216	**0.88**	**0.86**	**1.00**	**0.87**	**0.91**	**0.86**	**0.67**	**0.84**	**0.92**	**0.45**
22. Computer	16	0.92	0.91	1.00	0.83	0.90	0.72	0.58	0.90	0.94	0.61
23. Mechanical	64	0.89	0.89	1.00	0.91	0.93	0.89	0.70	0.86	0.91	0.48
24. Electrical	52	0.86	0.84	1.00	0.84	0.92	0.89	0.74	0.80	0.91	0.43
25. Other	13	0.85	0.91	1.00	0.83	0.88	0.76	0.52	0.84	0.92	0.43
26. Chemical	40	0.88	0.83	1.00	0.90	0.94	0.93	0.76	0.85	0.94	0.42
27. Civil	20	0.84	0.87	1.00	0.80	0.88	0.84	0.44	0.89	0.94	0.28
Mean		3.56	3.55	2.81	3.62	3.67	4.00	3.66	3.86	4.02	
Standard Deviation		0.62	0.70	1.00	0.62	0.53	0.60	0.86	0.64	0.57	

Note: Departments with fewer than five respondents were omitted; F1= Involvement, F2= Faculty Support, F3= Institutional Support, F4= Helpfulness, F5 = Diversity, F6= Integration, F7= Fit, F8= Engagement, F9= Importance.

getting a degree in their major are respected by most people," and "their future occupation makes an important contribution to society. *Institutional Support* had the lowest means (around 3.00) indicating that, on average, students across the racial/ethnic groups had mixed perceptions regarding support received from their institutions.

Differences by Race/Ethnicity

One-way analysis of variance (ANOVA) revealed no statistically significant race/ethnic differences in the theorized climate measures. In other words, each of the nine climate measures was similarly perceived across racial/ethnic groups. For example, *Institutional Support* had the lowest means (3.09 for Blacks; 2.87 for Hispanics; and 2.75 for Whites) which would be indicative of students across the ethnic groups having mixed perceptions on institutional support ($M \approx 3.0$), this measure also had the largest variances showing that the racial/ethnic group differences were minimal.

Suspecting that student perceptions on *Integration* (described earlier) might differ by institutions (e.g., FIU is an Hispanic serving institution; FAMU is a Historically Black college; and USF and UF are predominantly White), when we subset the data into Black and Hispanic samples and compare perceptions across institutions, there were no statistically significant differences in the climate measures including "Integration" among the four institutions.

Predictors of Intent to Leave

Of the nine data-driven factors from confirmatory factor analysis, three were predictive of *Intent to Leave* based on logistic regression analysis (not shown). *Faculty Support* referred to students' perceptions about the support and encouragement provided by faculty and staff to help students succeed academically, acquire research skills, and achieve professional goals. *Social Fit* referred to students' perceived social fit in the department. *Personal Agency and Peer Support* referred to a student's active involvement in the learning process and the support or atmosphere created by peers that encourage student's success. Figure 5.3 shows the items loading on each factor.

Prior to comparing the mean scores on the three factors, we established that a very high level of agreement existed among students at these institutions regarding these factors ($.85 < r_{wg} < 0.95$). As shown in figure 5.3, items with loading greater than 0.50 on a factor were retained thus the set of items was adequate in describing the factors. Considering the item

Figure 5.3 Means, Standard Deviations, and Item Loadings on Data-Driven Climate Factors

Factor/Item	Loading	Mean	SD
Factor 1: Faculty Support (α=.870)		3.56	.579
Faculty and staff are generally encouraging towards students.	.696	3.78	.870
Faculty and staff go out of their way to help students master the knowledge in their discipline.	.676	3.30	.957
Faculty are enthusiastic about teaching	.647	3.60	.925
Faculty and staff help students achieve professional goals.	.643	3.65	.836
Faculty and staff help students develop creative capacities.	.641	3.44	.896
Faculty and staff provide opportunities for students to work on research projects	.637	3.55	.897
Faculty and staff provide students with strong academic and professional role models	.612	3.68	.912
Faculty and staff are often available for students to see outside of regular office hours	.557	3.67	.993
Factor 2: Social Fit (α =.768)		3.49	.665
Within my department, I would easily be identified as "one of gang"	.615	3.32	1.09
I sometimes feel out of place.	.562	3.46	1.12
I do not feel "emotionally attached" to my department (R).	.562	3.17	1.09
I would be a good example of a student who represents my department's values	.512	3.66	.894
Factor 3: Personal Agency and Peer Support (α =.659)		3.81	.626
I have not yet "learned the ropes" of being student here (R).	.646	4.08	.990
Students are often too concerned with their own success to help each other	.571	3.45	1.07
I do not consider any of my fellow students as my friends (R).	.548	4.36	.966
I have not fully developed the appropriate skills and abilities to perform successfully as a student (R).	.545	3.91	1.06

Note: α = Cronbach's alpha; R= the item is reverse coded.

with the largest loading to be representative of the factor, *Faculty Support, Personal Agency and Peer Support* show means of approximately 4.0 implying students agreed that these two factors were important for their success while *Social Fit,* having a mean of approximately 3.0, indicates modest agreement with the statement regarding departmental fit. Additional analyses revealed no gender or racial/ethnic differences on how these factors predicted students' *Intent to Leave.*

Discussion

The pool of women and minority undergraduate students across the institutions of interest is one used to draw upon to increase the number of engineers needed to meet STEM workforce demands, a matter of concern to the nation. In this chapter we explored degree program climates associated with program efficacy for engineering students. Although we did not find statistically significant gender or racial/ethnic differences in students' perceptions of departmental climate associated with program efficacy, we established that *Faculty Support,* "*Personal Agency and Peer Support,* and *Fit* were predictive of *Intent to Leave,* a proxy for program efficacy in this study. How these factors relate to student program efficacy is discussed next, using qualitative data to provide insights on the relationship.

Institutional Support

Institutional support represents the support and encouragement provided by faculty and staff to students that assists them in their quest to succeed academically, acquire research skills, and achieve professional goals. It also includes resources such as laboratory, tutoring, advising, and other college or university-wide services. Students' perceptions of faculty support stem from classroom interactions, office hours, research and lab experiences, and mentoring/advising services. We discuss the extent to which students experienced each of these aspects of support.

Classroom interactions. Classroom interactions are the main forum for interacting with faculty. Students surveyed agreed that "faculty members were enthusiastic about teaching" but students were neutral about faculty "going out of their way to help them master the knowledge" or "helping them develop creative capacities." On the contrary, while almost all students interviewed believe professors are knowledgeable about the materials, they also believe some professors lacked the pedagogy to be effective as described in chapter four. Students mentioned that some professors lack interest in teaching in contrast to the enthusiasm they invest in their research. Some students also have difficulty understanding international professors due to their accent and sometimes, limited English proficiency. Despite issues related to faculty and classroom instruction, students did perceive some professors to be exceptionally enthusiastic, supportive, and valued their work: "...professors who are into their thing, into their element...willing to help out more...really into making sure that the students know."

Office hours. Office hours allow students to seek individual instruction and direction from their professors outside the classroom. Students surveyed agreed that faculty were often available for students to see outside of regular office hours, but few students made use of this time because office hours conflicted with their work schedules. Other students were uncomfortable with professors and preferred to seek help from peers or teaching assistants, viewing professors as a last resort. In one White male focus group, a student narrated how a professor stopped typing to pretend not to be inside the office. It may be that some professors are not able to cope with the after-class needs of students, perhaps because of other obligations, or are simply indifferent to students' needs. Some students explained how professors were much more personable, patient, willing to answer their questions and work through problems with them during office hours. In one White male focus group, students mentioned that while the majority of professors in their department are international in their backgrounds and struggle with English during lectures, these professors are much better at explaining concepts when speaking with individual students.

Advising and mentoring. Students reported that they value advising and mentoring; however, many students believe they lack opportunity to form such relationships with faculty. Students surveyed agree that faculty provide them with strong academic and professional role models. Students feel they must be proactive and take charge of their own academic experiences rather than expecting faculty or staff members to "hold their hand" throughout their time in the program. The amount of personal attention faculty can provide partly depends on the size of the department. In larger departments, advisors tend to be overloaded with students, making effective advising problematic. Students who did have a mentor identify with that individual because they perceive him/her to be similar to them in some way; for example, one student in the military talked about how he has a closer relationship with a professor because he also had military experience. Students believe it is incumbent on them to initiate the relationship, thus leaving advisor-advisee relations unstructured and contingent upon the personalities of students and faculty.

Staff support. Students interviewed agreed that staff were generally encouraging toward students and helped them achieve professional goals. Similarly, most students interviewed mentioned that staff members (e.g., office managers or department secretaries) were friendly, helpful, and took time to speak with them when they visited their offices. To students, simple day-to-day type of assistance, for instance, assisting with information about next semester's class schedule, gives them the impression the department is supportive and cares about them: "...a bunch of secretaries that always help me out with small tasks so I have much support."

Structural Support

Course sequencing. The sequence of courses within the Engineering curriculum tends to limit student's ability to complete course requirements. Many students interviewed believe their department did not care about them because if a course is only offered at one meeting time for only one semester per year, students must wait for an entire year to take the course. Students must be proactive in organizing their schedules long-term:

> So many times you'll need to take a class next semester and it's just not offered. Some classes are only offered once a year, some are only offered once every three years, so you really have to know you're going to be in engineering for the long haul, and you're gonna have to plan out your schedule and make sure that you can get all of these x, y and z requirements in. (FAMU-FSU White female student)

Engineering organizations. Organizations such as the *Institute of Electrical and Electronics Engineers (IEEE) provide forums for students to interact with* professors outside class time. With some notable exceptions, few professors participate in these organizations, however. When this is the case, students do not benefit from their help especially in terms of mentorship and advising.

Communication. Lack of communication limits the effectiveness of department structural support. Departments may provide resources and support services, but often students are not aware of them partly due to lack of communication about the existence of these services. Programs should make efforts to diversify methods of communicating important information to students while not overloading students with information in the form of e-mails, announcements, and flyers. Students may ignore a constant flow of information if they cannot filter what is most important. A student gave an excellent example of poor communication describing how the faculty member was appointed as the student's advisor without the faculty member's knowledge.

Personal Agency and Peer Support

Personal agency refers to students' active involvement in their learning process while peer support refers to the support or atmosphere created by peers that encourages student success. As noted by Bonous-Hammarth (2000), "student agency and peer group influences combine into a dynamic model to guide students successfully or unsuccessfully throughout their academic experiences" (p. 95). Students surveyed agreed that they had developed appropriate skills and abilities to perform successfully as students, "learned the ropes" of being students in the department, and that their peers were friendly and ready to help others succeed. Peer support is particularly helpful in large departments because in these environments students face challenges such as scheduling appointments with their advisor.

Faculty vs. peers: Whose help comes first? Students generally feel they must fend for themselves rather than rely on faculty, leading them to value peer support over faculty support. "I think the biggest thing that has got me through is just the friends I've had" (FAMU-FSU White male student). When asked where they go for help, almost all students, regardless of race or gender, responded that they would first go to other students before they would seek out help from administrators or faculty:

> If it's material wise I'll always go to my peers first to see if they understand it. Because most of the time the peers can explain it a lot better than the teachers...If I'm trying to figure out where I'm at with my grade I'll go to the teachers, ask them what can I do to boost my grade up in the class. (FAMU-FSU Hispanic female student)

While students seem to understand their peers better than their professors, when it comes to their grades and progress in class, they are willing to approach faculty who instruct their courses, but hesitant to approach those professors whom they do not know but who may be in their department.

Socio-emotional support. Relationships with other students help students feel that they are not alone in what they're going through and that other students are there for support. A student at UF described her relationships with other students:

> It's not like you're the only one studying until three o'clock in the morning.... You know that there's somebody in...that same class doing the same thing at that time, so you don't feel that you're the only one fighting so hard, that's it's ok to maybe have problems. So if you don't know anybody then that would be hard. (UF Hispanic female student)

Overwhelmingly, students mentioned friendships with other students helping them through the program. A student from FAMU captured this well, "Everybody works together...in my little group my friends that I've met.... They don't compete. It's more like trying to help each other because you know you can't always get it by yourself. You need help" (FAMU Black female student). Students generally define their interactions as collaborative rather than competitive.

Students also reported that friendly competition within groups was helpful. A white female student at UF said that her favorite aspect of the program and other students is "that it is competitive. That they keep me on my toes." A Hispanic male at FIU had a similar experience. Without other students, he claimed, "I wouldn't feel... I wouldn't have a competitive area. I do well when I'm competing. I don't do well when I'm all by myself." This creation of a friendly, competitive climate highlights students' ability to create their own climate.

Group work. Groups are formed in various ways and places. Frequently cited places to meet and form groups included the library, the fishbowl at USF, the atrium at FAMU-FSU, and other common spaces in the engineering areas of each of the campuses. Students recognize faces of peers from previous classes or peers who may be reading a textbook used in a given class and do not hesitate to approach them to form study groups. Thus, knowing each other personally is not a requirement for students to form study groups.

Students' perceptions on group work varied. A prevailing notion concerned how group work provided the opportunity to solve a problem together; no students viewed it as cheating. A few students noted that they believe the need to be prepared before they went to a group and used group work as a check for their homework. Students noted that most faculty members encouraged group work as it mimics what engineers do in industry: "they encourage us to work in groups," "working as a team which is part of what they [faculty] try to teach you in engineering," and "they [faculty] always preach, you know, when you get on the job site you're going to have to work as a team." Group work is also viewed as a source of emotional support to students. A UF White male explains, "It's good to have people that encourage and say well, just stick it out man, you know it will be alright." These qualitative findings support Goodman's (2002) finding that participating in study groups was both a source of academic and social support to female students. Other research shows that small group learning and peer support are crucial for increasing minority program efficacy in undergraduate engineering programs (Campbell, Jolly, Hoey, & Perlman, 2002; Pascarelli & Terenzini, 2005).

Social, academic, and emotional fit. Previous research on organizational climate points to the importance of the concept of fit in determining an individual's success in an organization. A great deal of research examines *person-environment (PE) fit*, the extent to which an individual's goals and values are congruent with those of the organization (Ostroff & Schulte, 2007) and *person-person (PP) fit*, defined as the extent to which the attributes (e.g., knowledge, skills, needs, perceptions, values, preferences, attitudes, and demographic characteristics) of an individual is similar to the same attributes of another individual or to that of other individuals in the organization in an aggregate form (Ostroff & Schulte, 2007). Following up the work of Tinto (1975, 1993), Astin (1984, 1999), and others this study found that student's perceived social, academic, and emotional fit is related to their success.

The four questions in the survey addressing fit captured the social aspect of fit. Students were neutral regarding the extent to which they perceived they identify as "one of the gang," feel out of place in the department, feel emotionally attached to their department, and feel that they represent their departments' values. Students were also similarly vague in interviews and focus groups when asked how they fit in their program. The overwhelming majority responded that they "fit in fine." "I don't feel outcast but I mean I don't feel like I'm unique at the same time" (UF White male student). Fit does take time. Students responding to the survey may perceive themselves as fitting in well because most students interviewed were in their third or fourth year in the program, "Well, I feel good about being here. The first year was super difficult... I didn't know any people... but as time passed that changed definitely" (FIU Hispanic male student). Academic fit is a primary concern for many students who may not personally rely on their peers for social or emotional support. "I am not at the bottom... like I said, some things I know really well: electronics, circuits, I know that really well, other things, don't ask me about it!" In general, students did not respond strongly for either fitting in or not; a common response was that they are in the middle or "half and half" as one student put it.

Gender and Racial Differences in Fit

The process of fitting in to the discipline is integral to understanding why engineering remains a predominately white and male field. Female and minority students interviewed recalled feeling more trepidation at fitting in than their non-minority counterparts especially in terms of academic fit. For example, one student thought the department would be "a bunch of geeks and stuff" (FAMU-FSU Black male student) and a Hispanic male in the same program worried that as an international student, he would not be able to keep up academically, "At first I thought I'm a minority in every way. I'm a woman. I'm Hispanic. I really didn't think I was going to fit in but eventually you realize you're not the only one" (FAMU-FSU Hispanic female). Fit is not static and it is perceived on different dimensions such as race, gender, and academic ability:

> I feel like I do fit because I have the credentials to be there, I can do the work just like everyone else, so in that aspect I guess I fit. But as far as

looking around and seeing people that look like me or probably think more like me, no, that aspect I don't fit. (UF Black female)

Minority students who fit academically through their prior preparation may not fit socially because of their race and/or gender. But minority students with an abundance of social capital through experiences with students of other races prior to entry into the program expressed less apprehension. One student attributed his comfort with fitting in to his high school experience: "I wasn't in...a state of shock when I saw people that don't look like me" (FIU Black male student). Having interacted with students from diverse racial/ethnic backgrounds while in high school, this student was not intimidated about being among non-Black students.

A chilly climate may not be overt, that is, a manifestation of blatant racism or sexism, but rather a matter of fitting a mold—either because one comes prepared for this atmosphere or because one gains it along the way. Female and minority students use words or phrases such as "adapt," "get used to," and "it's a slow process" to describe learning to fit in. "I fit in. I mean going into the class with all men, it's not like...You try to fit in...I don't see why any female should feel threatened or uncomfortable in any classroom" (USF White female student).

Interviews suggest minority students and female students are marked as different. They are either preferentially treated or they are said to be among the hardest workers in class, marking their behavior worthy of critical examination by their student colleagues, "It's kind of hard for other students to look at you as maybe a leader in a class because part you're Black and you're a female so they're like she's probably not that smart" (FAMU-FSU Black female student). Some students are reminded of their minority status when they are outside of their academic environment, "You get to the Swamp [the UF football stadium] and you're like, wow! There are only like 6% of us here, right?" (UF Black male student). The reminder that they are a minority in their larger institution may be an important negative experience, more so than overt racism or sexism. A "chilly" climate may manifest itself in very subtle ways, aligning with Bourdieu's (1977) contention that habitus is naturalized and that challenges facing women and minority students may not be visible to people in the majority. Interventions levied at creating greater gender and racial parity in engineering should consider the subtlety of experiences and the heterogeneity of students.

One student who changed majors several times to ensure he fits in highlights the agency students have in defining their experiences, "Once I got more involved with the [student] societies...I saw the opportunity to do more...I never spent one semester without being involved with clubs, societies...so I never had a concern for being scared, not fitting in at all" (FIU Hispanic male student). Many female and minority students credit their participation in student organizations with their fitting in to their programs. Chapter six explains that students who enter the program with less cultural capital rely on structured social networks such as student organizations to gain it, however a brief discussion is required here. The importance of student organizations that support women and minority students corroborates the contention of researchers who argue that collaborative and collegial departments are conducive to women and minority student success (Astin, 1984; Tinto, 1993). As we will see in Chapter six, compared to their White male counterparts, women and minority students did not feel their departments were supportive. As a result, they turned to other students, and through these relationships and student organizations they were able to create a collaborative and collegial environment for themselves. In effect, by surrounding themselves with students they relate to they were able to engineer their fit into the department.

Conclusion

Quantitative analyses suggest that there are no statistically significant differences in student perceptions of program climate associated with program efficacy for female and minority students; however, qualitative analyses suggest that departmental climate can be improved by providing more opportunities for student and faculty interaction to offset student perceptions of faculty involvement with their research. A strong recommendation here is the opportunity for undergraduate students to have access to faculty research projects. Student interviews and focus groups do not contradict or negate student survey results.

Findings in this chapter suggest that minority students are likely to rely on the social capital of other students rather than on faculty support. We suspect this is the case because of students' less powerful social and political economic positioning. This may reflect a strategy for students who feel marked as different to both gain the necessary social capital to succeed in the program and also to form friendships allowing them to experience camaraderie in a context both social and academic, that is, in student organizations such as NSBE that foster student identities as both engineers and members of social, racial and ethnic groups. Students are also likely to adapt to challenging engineering curriculum and pedagogy described in chapter four by working with their peers in the context of study groups.

Women and minorities may experience a "chilly" climate differently from that experienced by their majority student peers. Increasing the participation of female and minority students in STEM fields should be considered a multidimensional effort requiring various approaches. What appears to be superficially contradictory may in fact be an indication of existing complexities related to gender, race, and class. The mixed methods approach examines the complex issue of gender and race in the context of engineering degree climate. Findings underscore the need to triangulate data and to approach sensitive topics such as racism or sexism using several methodologies that give voice to individual actors while not sacrificing empirical rigor.

Notes

1. Florida State University Interactive University Data (http://www.flbog.org/resources/iud/ accessed: January 25, 2010) show the gender and racial/ethnic compositions were 20% female; 55% White, 26% Latino/Hispanics, 18% Black, and 7% Asian.
2. Unlike the theory-generating model found in exploratory factors analysis, CFA is a theory-testing model in which the researcher begins with an a priori hypothesis that specifies which variables are correlated with which factors and which factors are correlated with each other. The hypothesized relationships among the nine constructs and intent to leave were based on a strong theoretical foundation from prior research.

Chapter Six
Program Culture: How Departmental Values Facilitate Program Efficacy

Susan Chanderbhan Forde, Cynthia A. Grace, and Bridget A. Cotner

Introduction

Culture is a word we hear and see used in many contexts, political, educational, and social. "Changing the culture of Washington," became a mantra during the 2008 U.S. presidential election, and culture continues to be evoked in the economic crisis that has escalated in 2009. In this political context, we might interpret culture to mean "the way things are done." To many the word culture refers to "high culture," eliciting images of art museums, classical music, and theater. But what does the concept of culture mean to social science researchers focused on understanding women and minorities in science, specifically engineering departments?

A working definition of culture used in this study begins with concepts driven by research in industrial/organizational psychology referring to the values and goals of an organization, in this case undergraduate engineering programs. Our work is informed by an anthropological understanding of power drawn from political economy, as well as related beliefs that individuals possess agency making the development and maintenance of culture a dynamic process. Understanding culture as beliefs and behaviors adaptive in a given environment allows an examination of how engineering program culture impacts the attrition and retention of women and minority engineering undergraduate students.

Within the context of undergraduate engineering, beliefs and behaviors consist of what women and minority students as agents *think* as well as what these students *do*. For example, some women *think* they do identify personally as engineers while other women do not. Many women and minority students *do* behaviors and strategies such as creating study groups, building social support systems, and establishing mentors as described in chapter five. Students employ these adaptive strategies to facilitate their success in their engineering program. This definition is consistent with practice theory and elements of political economy described in chapter one. Some members of a culture, cultures, and organizations may have more power than others. We contend that women and underrepresented minority students have less power and cultural capital in a male dominated arena, engineering (Dirks, Eley, & Ortner, 1998). Women and minorities create alternative structures and forms of capital, demonstrating agency and self-efficacy. Alternative structures provide pathways for subsequent groups of underrepresented students.

This chapter seeks to understand the culture of engineering departments as articulated by women and minority engineering students and White male engineering students. We also spoke with faculty, deans, and administrators who contribute to departmental culture by setting expectations for success and have observed successful and unsuccessful strategies over time. This chapter analyzes interviews and focus groups with administrators, faculty, and students to understand the culture of undergraduate engineering programs at five universities in the state of Florida. Major themes from these analyses reveal that administrators and faculty members place a high value on research and teaching for undergraduates in engineering. Undergraduate students across institutions view faculty support and research as institutional priorities, but also discuss how a focus on research detracts from classroom instruction and mentorship.

Student strategies described in the chapter involve institutional and social forms of student support, and personal strategies employed by women and minority students. These strategies are forms of capital that are particularly important for students who come from disadvantaged backgrounds or who work against ideologies operating to devalue their capabilities and undermine identities as engineering students.

Program Culture and Program Efficacy

Program cultures with a strong student orientation provide support programs for students, stress collaboration among students rather than competition, emphasize collegiality not bureaucracy, and predict retention for all students, particularly underrepresented minority and women students (Berger, 2002; Braxton, 2002; Noel, Levitz, & Saluri, 1985; Pascarella & Terenzini, 2005; Smith et al., 1997). The BEST (Building Engineering and Science Talent) report (2004) identifies undergraduate programs with exemplary records of recruiting and retaining women and minorities in

STEM fields, that is, high levels of program efficacy, and notes that these program cultures share several common features including:

- placing equal priority on research *and* student learning
- emphasizing personal attention for students through the use of programs to promote faculty-student interaction
- valuing collaboration, particularly between students from diverse backgrounds
- providing students with professional and research experience, and commitment to constant evaluation of program progress in recruiting and retaining women and minority students

Research examining graduation rates of women from STEM programs identifies aspects of a supportive program culture important to the success of women enrolled in engineering, particularly when women are supported through positive relationships with advisors, mentors, and participation in study groups and groups targeted toward women in engineering (Brainard & Carlin, 1998; Goodman, et al., 2002). Aspects of program culture that negatively affect the retention of women include: a too competitive program culture (Goodman, Cunningham, Lachapelle, Thompson, Bittinger et al., 2002; Seymour & Hewitt, 1993) and a lack of emphasis on good teaching (Goodman et al., 2002).

With underrepresented minority students, an emphasis on support and collaboration is a key to their success in STEM undergraduate programs. In a comprehensive review of research and reports published since 1980 on minority recruitment and retention, May and Chubin (2003) identify several factors supporting minority students' success including having specific programs in place to sustain student success, particularly programs focused on academics. In addition, minority students fare better in program cultures promoting collaboration especially when programs use specific mechanisms including freshman orientation and structured study groups. Small group learning, peer support, programs and support staff providing assistance to minority students all increase minority student retention in undergraduate engineering programs (Campbell, Jolly, Hoey, & Perlman, 2002; Pascarella & Terenzini, 2005; Penick & Morning, 1983).

Informal peer groups composed of women and minority groups, as well as formal institutional supports including structured study groups, promote academic achievement. Goals and values espoused by faculty, administrators, and staff and how well these goals connect to programmatic efforts in student retention help in understanding how Florida engineering program culture promotes the retention of women and minority students. Understanding expectations held by these gatekeepers allows an understanding of program culture from the perspective of women and minority students.

Findings

Student, faculty, and administrator interviews were conducted in accordance with methods and analyses described in chapter one and employed in chapters three to seven. Students, faculty, and administrators have different values and goals concerning the importance of research and teaching, institutional support, student collaboration, and diversity, including the recruitment of women and minorities.

Analyses of the responses of administrators to questions concerning these matters reveal six themes across engineering departments:

1. valuing *research* (15 of 23 administrators, 65 %)
2. having *quality teaching* (15 of 23, 65 %)
3. enhancing the *doctoral program* (9 of 23, 39 %)
4. having gender and ethnic *diversity* among students (7 of 23, 30 %)
5. improving the engineering department's *ranking* (6 of 23, 26 %)
6. having a *collegial* department (6 of 23, 26 %)

Deans and program directors manage program infrastructure and, therefore, emphasize institutional goals. Administrators value aspects of engineering programs enhancing overall program prestige and continued departmental success.

Engineering faculty members emphasize two core values: (1) *research* (20 of 27 faculty members, 74 %) and (2) *quality teaching* (16 of 27, 59%). These values mirror prominent standards mentioned by administrators; nonetheless, faculty did not mention factors linked to departmental prestige including doctoral program development, diversity, department rankings, and collegiality.

Gaps in perceptions of institutional goals, values, and priorities (what is valued and emphasized in an institution) between administrators and faculty members may reflect differences in espoused values and enacted values. Research from I/O psychology notes that organizations may have differences between their *espoused* values (the stated values of the organization) and *enacted* values (the actions and practices of the organizations) (Ostroff, Kinick, & Tamkins, 2003). For faculty members, the goals of focusing on research and high quality teaching are enacted in their positions as a member of the engineering faculty—faculty members conduct research and teach students. Other goals espoused by administrators—enhancing the doctoral program, having student diversity, improving departmental ranking, and collegiality—are not explicitly mentioned by faculty in any of the participating programs. Prioritizing research and teaching at the expense of what can be termed "collective values," values that promote shared good will, suggests faculty are more concerned with their individual responsibilities and achievements as opposed to departmental and institutional goals.

Students' experiences of program culture are much different. Student values reflect adaptations students make to successfully navigate engineering departments. Students value:

1. faculty *support* for students (20 of 47 students, 43 %)
2. the importance of *collaboration* (18 of 47 students, 38 %)
3. the importance of *research* (13 of 47 students, 27 %)
4. applying knowledge to the *real world* (11 of 47 students, 22 %)

Students rely on support from faculty and their peers to learn how to meet expectations. Research and the application of knowledge to the real world are valued as links between their initial interest in engineering and the theory and problem-solving skills valued by classroom pedagogy, both described in chapter four.

Faculty and administrations establish department culture through their longevity and institutional memory. Students enter an engineering program with an established culture and react to their new environment, often adapting to emergent norms and expectations and sharing in or diverging from the values and goals of institutional gatekeepers. For this reason, each section begins by examining the values and beliefs of faculty and administrators followed by those of students.

Teaching vs. Research

The importance of *research* and *quality teaching* are prevalent themes in interviews with both administrators and faculty members across institutions. An emphasis is now currently placed on enhancing the research agenda in each of the engineering departments included in this study. Faculty members are expected to pursue scholarly achievements, outside funding, tenure and promotion, and program prestige. With the need for more graduate students to facilitate faculty research endeavors, the theme of *enhancing the doctoral program* is linked with the value of research. Similarly, improving an *engineering department's ranking* is described by UF faculty and USF administrators as being an outcome of research prestige. Several students attribute the poor instruction they receive in most classes to the tension between teaching and research experienced by faculty.

Students admire faculty research achievements but believe that time spent on research takes away from time developing effective pedagogical strategies discussed in chapter four. In addition, few undergraduate students are given opportunities to directly participate in research activities with faculty and graduate students.

Faculty and Administrator Perspectives

Research. Research is important to the individual goals of faculty and collective goals of the university. Funding from on-going research projects supports faculty through the summer and brings regional and national prestige and additional outside funding to the university.

> The organization rewards [research] because I suppose of the overhead return and the [fact that] the university also gets recognized for the high research component and from the individual's viewpoint they also get money...for their summer support and also [the research] strengthens their resume so they become more marketable. So in my opinion when you put those two components in there it's a no-brainer. (USF administrator)

Quality teaching. Teaching does not bring the same level of prestige or financial reward, but is essential to the instructional goals of the university—each of the universities is expected according to its mission to ensure high quality instruction at the undergraduate level. FAMU-FSU and FIU faculty members and FIU administrators focus on maintaining balance between research and teaching by not treating one as more important than the other. Through teaching, "We want to make sure we transfer knowledge so they don't have to reinvent the wheel again" and through research, "We want them to know the technology moves so fast so you've got to keep up" (FIU administrator). Faculty members at FIU agree with their administrators about the need to balance teaching and research, "We have to set a balance between teaching and research, because without teaching the university doesn't exist.... We try to keep the high quality of teaching and at the same time pushing research" (FIU faculty). Faculty members across the institutions with the exception of USF describe maintaining a balance between research and teaching priorities.

USF faculty members express concern about tension between faculty who engage in research and faculty who focus on teaching:

> ...There are faculty who essentially focus all of their attention on the research work. They do a very good job with it, but I get the feeling we often do it at the expense of the teaching responsibilities and I question the

overall productivity of that approach. Is it really helping the department and the college? (USF faculty)

Other USF faculty members believe the focus should be on undergraduate students: "Our final product is our undergraduates so we need to take care of them" (USF faculty). Building the graduate program requires an emphasis on research; however, most USF engineering faculty feel that producing undergraduate engineers is their principal concern.

Administrators agree high quality undergraduate engineering education is a major priority. Superior research as a core value promotes quality education beyond teaching, and demands a balance between undergraduate and graduate education. An administrator at UF explained, "I don't want that to happen. We are a large public university. We have an obligation to the undergraduate student body to give them the highest quality education." Some administrators fear that programs may over-emphasize graduate education to the detriment of undergraduate programs.

Doctoral program. FAMU-FSU and UF administrators seek to improve the doctoral program at those institutions as part of the value placed on research. An administrator describes a "seamless" connection between research goals, teaching goals, and the graduate program saying:

> You have undergraduate students and you have graduate students who are often times their TAs...who are a very important part of their educational experience for undergraduates who come here because of the quality of the faculty and the research that they are doing. So here you have the faculty who are trying to better the practice of graduate students, faculty [unclear] in the classroom and the graduate students are also in through the TA work, influencing undergraduates so no piece stands by itself. I mean there's a real connectivity between the three pieces of what we do and I think that's a core value, core value for the college. (UF administrator)

UF administrators point out the importance of research and undergraduate teaching, but also highlight the production of graduate students as a primary goal. "We do undergraduate education...and I think we do it well...I would say our primary focus right now is research and PhD production" (UF administrator). Graduate student production is also a priority at FAMU-FSU, "We have a good number of undergraduate[s]...and we're almost to the saturation level with undergraduate population. But we are still short of graduate population" (FAMU-FSU administrator). In contrast to administrators, faculty members at FAMU-FSU and UF and their peers at other institutions did not mention graduate program enhancement as a specific goal. Clearly this issue can be seen as an institutional concern rather than a personal concern of most faculty members. Like valuing high quality teaching, valuing graduate education might be regarded as an espoused rather than an enacted value. Several faculty mentioned their stake in graduate program recruitment as fairly minimal, a function of their primary responsibility in undergraduate instruction. Nonetheless, as research funds come into a department, faculty need graduate students to assist with projects, "Without graduate population we can't do the research. Without graduate population we can't bring research....Right now it's our aim to increase the graduate population especially at the Ph.D. level" (FAMU-FSU administrator) Graduate students benefit from this arrangement by receiving practical experience and mentorship by the faculty.

Engineering departments desiring to increase graduate student enrollment may recruit higher caliber graduate students with interests and expertise in specific areas related to faculty research agendas. Administrators at FAMU-FSU and UF view graduate students as important resources to fill research and teaching assistant positions, obtain research funding, and successfully complete research projects, thus enhancing the overall prestige of the department.

Ranking. UF faculty members and USF administrators prioritize the importance of having a high ranking engineering program. UF faculty members seek national recognition as a top five research department, a goal specifically linked to the research program and production of graduate students. USF is not a nationally recognized university in engineering but is not short on aspirations at the administrative level, "The university has set goals to be in the top 50 [research universities] and so that has of course trickled down to the college as well as the department so that's what we aspire to as well" (USF administrator). USF faculty did not mention enhancing the national ranking of the engineering program as a value. This may also be an example of espoused values in contrast with enacted values. Faculty engage in research to meet personal goals of obtaining research support and enhancing their bids for tenure and promotion as well as their competitiveness on the larger job market. Administrators focus on marketing the engineering program through touting program rankings and engaging in university politics and other efforts to promote the university.

Administrators and faculty members across institutions are united on the importance of research and teaching in engineering. Maintaining balance between research and teaching is emphasized at FAMU-FSU, FIU, and USF. USF faculty members do perceive tension among faculty members who invest their energy in research in contrast to those who teach.

UF faculty value both research and teaching, similar to other institutions, but administrators and faculty affirm the importance of undergraduate teaching while highlighting the importance of increasing the emphasis on graduate studies. Prior research finds that institutions emphasizing *both* research and teaching are more successful at recruiting and retaining minority students (BEST, 2004). Administrators at USF and faculty at UF link a successful research program to enhancing the doctoral program and increasing departmental ranking nationally.

Student Perspectives

Across institutions, 27% of students interviewed believe in the high value administrators and faculty members place on research. Despite their great respect for professors who are proficient in their research, students believe an emphasis on research detracts from teaching and student support, "I think that there are very few professors who honestly care about what we learn. I think that our school is really focused on research and bringing in money and grants and things like that" (FAMU-FSU Hispanic female student).

This focus on research prompts students to say that faculty "need to focus more on their students instead of just their research" (FAMU-FSU Black male student). Students also make a sharp distinction between research faculty and teaching faculty. Research faculty are not considered to be teachers. Instead, "they're people who bring money into the school because they can do research or they do things and…they don't know how to teach" (UF White male student). Students believe professors do not put much effort into preparing for their classes, blaming their focus on research for the poor pedagogy described in chapter four:

> A lot of the professors seemed sort of arrogant. They really relied on their TAs to grade everything and they were just there for research purposes…they'd put up a Power Point slide and just breeze through the material. (UF White male student)

Students also report that most professors are not helpful during office hours: "Only one of my professors was actually very helpful. When I'd gone to their office hours, everybody else just fed me the same rhetoric that they had given in class and I ended up at square one" (UF White male student). Students wish to feel comfortable with their professors, because they have so many questions:

> …I mean no teacher's really 100% focused on their class. The teachers are not really there for you to have that type of relationship where you can come to them and say I have a serious problem, I don't understand this. You

know they're kind of like, okay, well, that's not my problem right now. I mean there's a few that care but they don't really care about how much we learn. They just care about, "I have to give this test and I have to give you your grade so you need to do this test" not, "Are you really learning this material?" (FAMU-FSU Black male student)

This student emphasizes the importance of internalizing materials presented in class.

FIU students believe graduate students are given priority over undergraduate students "They're trying to graduate more and more master's degrees. They're interested in the research program. I don't think they're really including the undergraduate people" (FIU Hispanic female student). FIU students interviewed do not believe faculty research detracts from teaching quality, but research does affect faculty availability. Largely because faculty members are engaged in heavy commitments to their research agendas, all students report that mentorship is missing in their programs:

> I mean if you have an issue with your grading or the classes, some of the professors it's hard to get in contact with them....They go to conferences, they're in the lab. This is an engineering center, so there's a lot of research going on here and the professors here are involved with that research. (FIU Hispanic male student)

Although students say having a mentor is important, only one interviewee (an Asian male student) reports actually having a mentor. Unfortunately, *no* female students report having a mentor—these are the students who would benefit most from mentoring. Furthermore, none of the institutions in the study provides formal mentoring programs, instead relying on the pre-college transition support described in chapter five and on whatever informal relationships between students and faculty might develop.

Importance of faculty support. Despite students' negative view of the emphasis on research at their institution, 43% of students interviewed describe their programs as places where their learning was supported by faculty. Students view faculty as supportive, whether by being available to students to provide assistance, holding high expectations for students, or valuing teaching. Women and minority students express their appreciation of faculty support and encouragement. For example, an FIU student discusses faculty members' supportiveness in terms of their availability to assist students saying, "They have all managed to maintain their office hours. There's not one of them that wouldn't drop what they're doing just to answer a couple of questions" (FIU Hispanic male student). Another student describes the encouragement she received from a particular faculty member because of the high expectations he held for students saying:

> He really understands students, and he will not let you make excuses for anything you do... So it makes you feel like you can do more you know, he'll push you to do more and more... to be the best you can be. (UF Hispanic female student)

The extent to which students believe faculty members are supportive varied across schools. Students at UF and FIU mention different forms of faculty support and USF and FAMU-FSU students are more conflicted about faculty support. A student at FAMU-FSU notes that a major source of encouragement and support for her is the availability of faculty members and their investment of time in students to help them prepare for exams:

> The professors, like I said, are always willing to sit down and talk to you... [and] every semester before the fundamentals of engineering exam civil and environmental professors will come out and spend their evenings tutoring students. (FAMU-FSU White female student)

In contrast, another FAMU-FSU student expresses his belief that faculty members do not support students beyond very narrowly circumscribed limits:

> See a lot of this stuff I think that it's left up to us to find or to figure out. I think as a whole the teachers just teach their class. They don't really tell you about the department or [that] this is what you have to do to get from the starting to the ending. It's like they just [say] this is my class I'm going to teach this. I mean that's how I feel, like a lot of it is left up to the students to discover on their own. Most of the stuff I find out, I find out from my friends. Like there's an exam [the Fundamentals of Engineering exam] we have to take and you're supposed to take that closer to when you graduate. (FAMU-FSU White male student)

Faculty members and administrators place a high priority on research and teaching and often struggle to balance these two goals. Students perceive the emphasis that departments place on research and believe that this emphasis results in poorer quality teaching. Despite students' negative comments about how research impacts the quality of instruction, students describe the support they receive from faculty members in mostly positive terms.

Importance of collaboration. Many students also talk about the emphasis on collaboration among students in their programs, generally because of its importance to the real world of engineering. A culture that emphasizes collaboration during classroom instruction is more prevalent at FIU than the other schools in the study, with a majority of students there noting that collaboration is emphasized by faculty members. This is likely due in part to FIU's particular population and context. As a Hispanic-serving institution, FIU draws from its proximity to Little Havana and the south Miami area. Hispanic students may prefer collaborative learning situations (Irvine & York, 1995).

FIU, FAMU-FSU, and UF students note that faculty promoted collaboration because it would be important to their future work as engineers: "...a lot of the classes foster collaboration, especially all the engineering courses, it's always collaborative because we're always taught we're going to be working in teams...and you have to do stuff to help other people" (FIU Black male student). Similarly, application to industry was broached when group work for homework was discussed, "... [you can do] homework in three hours instead of doing it in ten, and I don't feel that is cheating; it's just working as a team which is part of what they try teach you in engineering. You have to work as a team" (UF Hispanic female student). While discussing group work a student compares it to a support group:

> ...it's like a support group...Things like that really, really help you and getting help for homework, getting help for exams, studying together, pulling each other through engineering...which is what helps. (UF Hispanic female student)

Despite the fact that faculty members emphasize collaboration and structure assignments to ensure that teamwork occurs, students themselves don't always share this orientation. Often, students note that despite an emphasis on collaboration by faculty, students themselves engage in competitive behaviors. This is the case even at FIU where cooperation is seen as a dominant value of the program culture, yet a student recounts, "...it can get somewhat competitive and it's usually with the people that have the higher grades" and noted that this focus on competition "comes from the students" (FIU Hispanic male student). Another student, who self identified his ethnicity as 'Other' said he believes that, "...the competitiveness comes from the individual so if you want to be on top of the class and you want to be number one [you must compete]" (FIU male student).

At FAMU-FSU and UF, where collaboration is not perceived as important in comparison with FIU, students also describe a disconnect between faculty priorities and student actions,

> I think the professors do more...collaborative work with each other whereas the students create that competitive atmosphere. It's not everybody,

but there's some you know who want to be the best in the class, so they work towards competing with each other, and then naturally if you have that competitor spirit in you, you want to compete. So it's not like the teachers are saying who can be the best, for the most part the professors say work together. (FAMU-FSU Black male student)

Similarly, another student notes that professors do emphasize collaboration but that for various reasons students, including himself, didn't always work in groups:

> Well, they [the professors] tell you [you] can work in groups. But most people I think don't work in groups... you have other things like schedules and times and things like that. For me it's just that I'm so used to working on things on my own. I can work in groups, but I would prefer to do it on my own. I really feel I can. That's just how I am. (FAMU-FSU Black male student)

When asked if competition is encouraged among students, one student says, "... no I think it's just there [are] people [who] are competitive and that's just how we're going to be" (UF White female student). Whereas no students at USF mentioned collaboration as a value of the program culture, a student from UF described other students as very competitive:

> Some students are passionate but sometimes it goes to an extreme, where that's all they care about. So at that point then they don't care about you, they don't want, to have a friendship with you. If you find something that they're interested in, then it'll benefit both parties. If not then, you know, some students will basically have nothing to do with you. For example, this 13 student class, it's sort of, we're all competing against each other. That's how I feel. So does that lead to friendship? I would say no, which is a shame. (UF White male student)

Thus, the competitive atmosphere reported at UF stems from the students' competition with other students.

Program cultures characterized by collaboration have been identified by a large body of previous research (e.g., Pascarella & Terenzini & BEST, 2004) as important to the retention of undergraduate students in general and underrepresented groups in STEM degrees in particular. It was clear in student interviews that students understood that faculty prioritized collaboration; however, students did not always respond in expected ways. Analyses were conducted to determine if there are differences by gender and race for the importance of collaboration. These findings are presented later in the chapter.

Real world application of knowledge. Engineering careers involve practicing problem solving and applying knowledge acquired during the undergraduate experience—a five year program—to problems of immense complexity in the fields of both civil and electrical/computing engineering. While relating practical knowledge to the real world is discussed by fewer students (22%) in the study than collaboration and research, it is nevertheless an important theme. Students note that this is an important value in their programs. This was especially true for UF students (38%). Students who claim their institutions value applying knowledge to the real world express that faculty constantly attempt to link concepts embedded in the curriculum to their application in real world settings. A student describes this:

> And most all the professors that I've had will try to point out... that this is used here. When you graduate, this is what you're going to apply, and not just, you know, repeating something that a TA could, but trying to give us the whole picture. And then of course you know pointing out that that once we get out... we won't have to worry about... scrambling for a job (UF White male student).

Another student notes that she believes the department's priority is career preparation, "They're focused. They're there to teach you how to be an engineer... to make you the best engineer you can be" (UF White female student). A FAMU-FSU student discusses the department's increasing emphasis on applying concepts learned to the real world, "Something you're seeing more and more is application not just learning it, but learning how to use it in the real world. Like [what] we're doing in the senior design project, that's why they changed it" (FAMU-FSU White male student).

A majority of USF students interviewed articulate applying knowledge to the real world as a goal of program culture. USF wants students "To graduate with the knowledge and work and do what's right for the society and your client" (USF Hispanic White male student). Another USF student describes a professor's approach to illustrate his view that the priority at USF is preparation for the real world:

> [H]e says this is how it's going to be. If you all can't handle it in my class, then you're not going to handle it in the real world. I mean they try to get you ready for having to think about one thing and trying to apply it to something. That's what he does. He gets a bunch of different things together, could be from all areas of engineering and just pumps it into one question. He says, "Figure it out. That's what you're going to do out there." There's so many different obstacles in the real world that a textbook can't teach you. And I think the professors here; they don't baby us (USF White male student).

Overall, students articulate that their programs emphasize preparing them for the reality of engineering practice. Students describe faculty members as emphasizing applying concepts in realistic ways. Students view this emphasis on applying knowledge to the real world in mostly positive ways because they believe they are being prepared for the challenges they will face as engineers.

Gender and Racial Ethnic Differences
Faculty and Administrator Perspectives

Diversity. Increasing the *diversity* of the engineering student population is a theme among administrators at UF, USF, and FAMU-FSU but not among faculty members across institutions or administrators at FIU. At FIU, an administrator explains why diversity is *not* an issue on the FIU campus:

> We are I think in the nation the college that graduates more minority engineers than anybody else in the country. So we're number one at that. And so that's not a problem. And I think we are number three in terms of women. So we do fairly well in that sense. But again it's all about I guess location. How can you be in Miami and not be diverse? (Hispanic male administrator)

Administrators at FIU recognize the importance of having a diverse student population in their engineering program and indicate that the level of diversity in their program is a source of pride for them. Tapping into the local, national, and international student pool is one of the methods used at FIU, but for institutions that are not located in areas that are ethnically diverse, a priority has been placed on increasing the diversity of students and faculty members by administrators across institutions. Having a diverse engineering student body is also perceived as a necessity to grow in the future:

> We're always trying to increase the numbers of minority faculty and students, again engineering it's a non-trivial thing but it's something we have to do because that's where the pools of talent are gonna be moving forward so we can't rely on overseas students any more 'cause there's so much competition for them and there's all kinds of reasons why that's not as easy as it used to be. So diversity is a big one. (UF administrator).

At FAMU-FSU, increasing the diversity of the students through greater numbers of women and minority students in the engineering program is also seen as a goal by two administrators. With the merging of a historically black university, FAMU, and a traditional university, FSU, administrators in particular se an important value in continuing to increase the diversity of the engineering students graduating from the joint program. FAMU-FSU offers a unique structure for a college of engineering in that students and faculty members from a historically Black university and a predominantly White university (FAMU and FSU respectively) are members of one joint engineering program, although individual faculty members are affiliated with one faculty or another.

Collegiality. Perhaps because of the unique structure of their institution, administrators at FAMU-FSU also talk about providing a positive environment through improved collegiality. FAMU-FSU administrators describe the importance of having a collegial engineering department. As one administrator points out, "...the one thing people talk about a lot that they seem to value would be that we all get along. I'm not sure I've completely seen that in action, but that's what they talk about." Administrators there point out that, "We're a college here but two universities" (FAMU-FSU administrator). Because these programs were folded into one joint program, administrators at FAMU-FSU are focused on creating a cohesive college of engineering that focused on research, teaching, and getting along while enhancing their doctoral program.

While administrators highlight the importance of having a diverse undergraduate engineering student population and faculty, faculty members across all institutions did not mention diversity as a value of the department. The omission of this value by faculty members is an indicator that even as administrators support the idea of increasing diversity, strategies to promote diversity are not being enacted by the department. Furthermore, department faculty members do not determine undergraduate admissions and have little control over the process.

Student Perspectives

Research on organizational cultures (e.g., Ostroff, Kinicki, and Tamkins, 2003) indicates that it is possible for organizational cultures to contain *subcultures* displaying different values than other subcultures in the organization. Women and minority students may create subcultures in order to best adapt to being marginalized within engineering programs. Accordingly, student experiences were analyzed with respect to how they reflect institutional priorities. Women and minority students were compared to White males to examine how their relative experiences reflected differences in institutional priorities. A comparison of women and minority priorities to White male priorities revealed differences in perceptions of institutional support for students and the importance of collaboration.

Institutional support for students. This is defined as support provided through student organizations, department staff, flexible course scheduling, and provision of social activities. Twice as many White males as women and minority students say that institutional support for students is valued by members of their institution (23% vs. 11%). Furthermore, 14% of women and minority students cite a lack of institutional support such as poor advising, lack of communication about programs that support students, and communicating to students that they were not valued, as characteristic of their school cultures while none of the White males interviewed did so. One student describes a perceived lack of caring by the advisory team saying, "It bothers me that it seems like our advisors don't really care because they're kind of in/out" (UF Black female student). This student clearly feels dismissed by her advisor which she may internalize as not feeling important or supported.

Women and minority students' negative perception of institutional support explains why they rely heavily on women and minority student organizations and, when they are available, other programs geared to support the success of minority students. One example is assistance provided to freshmen and sophomores by student organizations. A student at FIU discusses the support she received from a student organization, the Society for Hispanic Professional Engineers (SHPE) to which she belongs. She states:

> Well,... I think we just provide a good atmosphere to just have fun and at the same time do productive things and get to know your peers....When we have a meeting we want to make it fun where you're gonna be working but at the same time you're going to get to know these people who are really cool and they're in your major and they're gonna help you. Because one of the things we say at the meeting is, "Hey, if you need help call me, I'll give you my email, if you need help with calculus especially for freshmen and sophomores. I can be your tutor for free; you don't need to go tutoring, which is $20.00 or something." (UF Hispanic female student)

This fits with prior research (e.g., Goodman et al., 2002) that shows that participation in study groups and social organizations functions not only as a source of academic support for women students, but also as a source of social support, i.e., both social and academic capital.

Supporting previous research, women and minority students (47%) interviewed also describe the *importance of collaboration*. In contrast, only

14% of White males interviewed viewed collaboration as an important value. One student expressed the value of camaraderie,

> The student camaraderie has been very, very valuable. I think that's the main thing that gets a lot of people to succeed in the department. Students all know each other and don't have a problem speaking to one another [and] asking for help. (FAMU-FSU White female student)

Female and minority students articulate that the nature of the engineering profession emphasizes collaboration. A FAMU-FSU student says, "It's always collaborative because we're always taught we're going to be working in teams. We're always going to be working with teams. And you have to do stuff to help other people." (FAMU-FSU Black male student)

White males do not indicate that collaboration is a value, instead noting that not everyone likes to work in groups. According to one White male student at FAMU-FSU, "[The professors] encourage us to work in groups, like they'll even walk by us sometimes and ask a lot of questions" the student continues, "There are some loners...that do the work on their own."

Administrators across the four schools articulate the importance of diversity, of recruiting and retaining both women and underrepresented minority students. However, with the exception of a program at one school, the Successful Transition through Enhanced Preparation for Undergraduate Programs (STEP UP) at the University of Florida, there appear to be few formal mechanisms in place to support the recruitment and retention these students. In addition, faculty, those who interact most closely with students, did not discuss diversity as a value of their respective departments. The lack of a clear systematic commitment to recruiting and retaining minority students and concomitant policies perhaps accounts for the perceptions of women and underrepresented minority students of their institutions. This is clearly a policy issue that is larger than engineering departments and must be addressed at multiple levels within the university. In comparison to White male students, female and minority students report that they receive lower levels of institutional support. Student collaboration may reflect one of the strategies that have evolved for coping with the lower levels of support they receive and the lower levels of cultural capital that they bring to engineering programs.

Discussion and Conclusion

Understanding how faculty and administrators on the one hand and students on the other view program culture, implies identifying the meanings of words used by these individuals in talking about their departments.

The question of central importance to our work is how these views are different or overlap with positions held by women and minority students. While women and minority students may not perceive faculty as less supportive than their White male counterparts characterize them, they did believe institutional support overall was lacking. In contrast, White males report institutional support from their standpoint is satisfactory. Based on administrator and faculty interviews, only administrators espouse the importance of increasing diversity within engineering departments. Without formal mechanisms of support (e.g., mentoring programs, research experiences) that research shows leads to the success of women and minority students, an increase in diversity of undergraduate engineering students and faculty members will be difficult to achieve. In addition, the quality of advising likely affects minority and women students and White males differently. Because underrepresented students entering undergraduate engineering programs may possess less cultural and symbolic capital to successfully navigate these environments, the weaknesses in some areas of institutional support likely affect them more negatively.

One strategy that women and minority students adopt to cope with poor institutional support is depending upon each other as resources. Students in the study report utilizing student organizations such as the National Society of Black Engineers (NSBE), the Society of Hispanic Professional Engineers (SHPE), and the Society of Women Engineers (SWE). When they are available, students also report that programs aimed at minority students such as the STEP UP program at the University of Florida constitute an important source of support.

Students' discussion of collaboration reflects how important this dimension of their experience is to helping them succeed. In this study, female and minority students recognize and respond positively to faculty members' emphasis on collaboration. In contrast, White male students are not receptive to such messages. Since collaborating in various ways has been identified as an aspect of program cultures that supports female and minority student success, it is important that engineering programs institute formal mechanisms such as structured study groups and freshman orientation to ensure collaboration occurs (May & Chubin, 2003).

The political and economic position of most contemporary state universities is one of increasing dependence on external funding, particularly from research grants, as state aid is steadily reduced. Given the importance of research to the financial future of most state universities, the emphasis on research found in the programs in this study will continue as an important priority for many undergraduate STEM programs as long as the economy makes a slow recovery. However, the critical shortages of women and minorities in STEM fields are also a pressing issue for universities and the nation as a whole. Thus, engineering departments must find ways of balancing research priorities while creating department cultures that are supportive of the success of women and minority students. Many students see their universities as providing little institutional support. Research shows that the persistence of women and minority students is dependent on such support (Goodman et al., 2002; May & Chubin, 2003). Departments should consider creating more formal mechanisms (e.g., mentoring programs, research apprenticeships) specifically for women and minority students. These strategies assist underrepresented students in building supportive relationships with faculty, and provide the cultural capital needed to navigate undergraduate STEM programs. Providing faculty members with students to assist them with research is an additional strategy that supports institutional priorities on producing research.

Women and minority students develop practical and workable strategies including depending on women and minority student organizations for academic and vocational assistance and social support. While students find informal ways to successfully complete their degrees, research on programs that are efficacious in supporting the success of women and minority students tells us that clear institutional commitments, including formal mechanisms of support, is critical to the success of women and minority students in STEM programs.

Chapter Seven
Making the Transition: The Two-to Four-Year Institution Transfer Experience

Cassandra Workman Whaler and Jason E. Miller

Introduction

Community colleges are emerging as important institutions for the production of science, technology, engineering, and mathematics (STEM) graduates. More than 40% of all undergraduate engineering majors nationally attended community colleges at some point in their academic careers (Tsapogas, 2004). Most importantly, community colleges are used by greater numbers of minority and female students, older students and students from lower socioeconomic strata; all underrepresented in engineering. The demographics of community college students are certainly not static—a rise in attendance by traditionally aged students and a decline in attendance by older students is predicted to continue (Bryant, 2001). Since women and underrepresented minorities attend community colleges in high numbers, the community college pathway to four-year institutions is an important context for research. It is likely this trend will continue during lean economic times as community colleges are a less expensive option in comparison with other kinds of institutions. As such, researchers and STEM educators are calling for more research into understanding the transfer experience for female and underrepresented minority students.

The purpose of this chapter is to understand how community colleges in Florida contribute to the production of STEM graduates on engineering pathways from community college to degree completion. We examine student focus group and interview data gathered from current community college students and from transfer students attending the university-based engineering programs of interest. Community college students who were interviewed attend "feeder schools" located in close proximity to the five universities that were the focus of the study. We also analyze data from workshops conducted with minority students from the University of South Florida, the majority of whom are engineering students. These data are then used to outline the pathways community college students take in attaining a STEM degree and the obstacles they encounter as they transition from community colleges to four-year universities.

Race/Ethnicity

While a higher percentage of White students earn four year Baccalaureate degrees, more Black and Hispanic students earn either Associate degrees or Certificates from community colleges than their White counterparts. Figure 7.1 demonstrates the differences between race and ethnicity of students enrolled in community colleges and those attending four year universities.

Within Florida, there is a wide range of variability in the geographic distribution of underrepresented minorities in four year institutions (see figure 7.2).

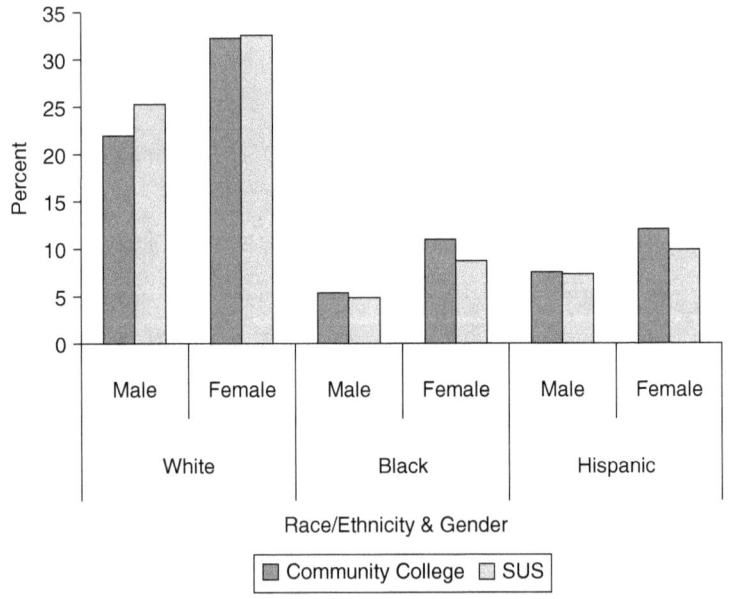

Figure 7.1   Total Florida Enrollment, 2006–2007

Source: Adapted by the authors from the Florida Board of Governors Interactive University Database.

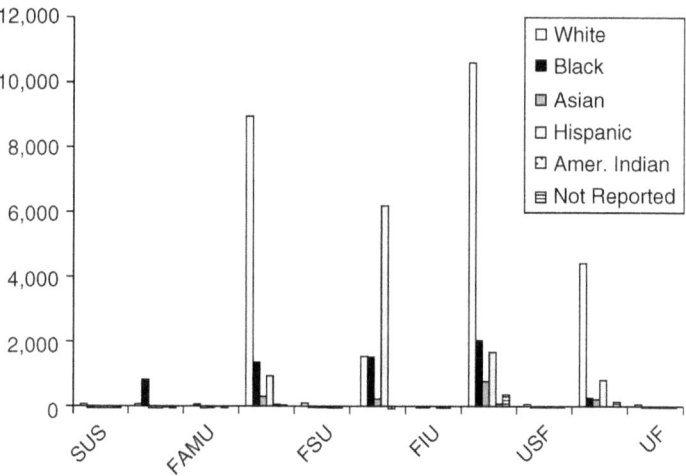

Figure 7.2 2006 Florida Community College System Transfer Students by Race

Source: Adapted by the authors from the Florida Board of Governors Interactive University Database.

For example, at Florida A&M (FAMU) 85.1% of transfer students are Black while at the University of Florida (UF) only 4.6% are Black and 71.9% are White. At Florida International University (FIU), 62.3% are Hispanic. The second highest percentage of Hispanics (14%) attend UF. However, an examination of the distribution of transfer students enrolled in Florida State University System (SUS) institutions shows a majority of transfer students are White. During 2006, in Florida SUS institutions, 62.2% of transfer students were White, 12.5% Black and 17.3% Hispanic, suggesting that while large numbers of minority students attend community colleges, a majority are not transferring to SUS institutions; thus, even fewer students are entering engineering.

Women Students

Women are the majority (58%) of the community college demographic (Phillippe & Patton,1999). The percentages of female transfer students at SUS institutions mirror this distribution. The Florida SUS average for female transfer students is 59.4%, with a high of 61.9% at USF and a low of 47.3% at UF, arguably the most selective of the institutions included in this research. UF is the sole university with a male majority among transfer students (see figure 7.3).

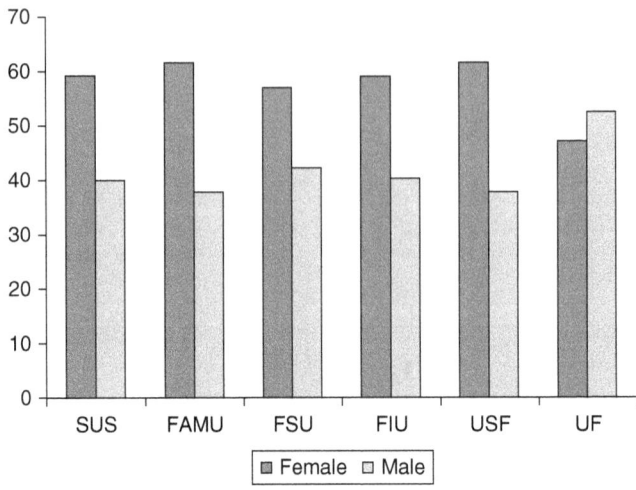

Figure 7.3 2006 Florida State University System
Source: Adapted by the authors from the Florida Board of Governors Interactive University Database.

Older students attending community colleges are most likely to be employed and attend community colleges part-time (Bryant, 2001). It appears, then, that female students make the transfer from community colleges into four-year programs with more frequency than their male counterparts and yet they are less likely to choose STEM fields, engineering in particular.

Students of Lower Socioeconomic Status (SES)

Community college transfer students in Florida receive more Pell Grant awards (federal grants based on economic need) than their four year degree earning counterparts regardless of ethnicity (see figure 7.4).

In contrast to First Time in College (FTIC) students, among students of all ethnicities, there was a 21% increase in the percentage of Pell Grant awards among Community College Transfer Students (CCT) from 1999 to 2003 (Borman et al, 2007). Because Pell grants are based on economic need, they can serve as a proxy indicator of socioeconomic status. Not surprisingly, according to national data, the 1998 median incomes for families with 18- to 24-year-old dependents were: Blacks, $27,000, Hispanics, $26,000, Whites, $59,000, and Asians, $50,000 (Mortenson, 1999).

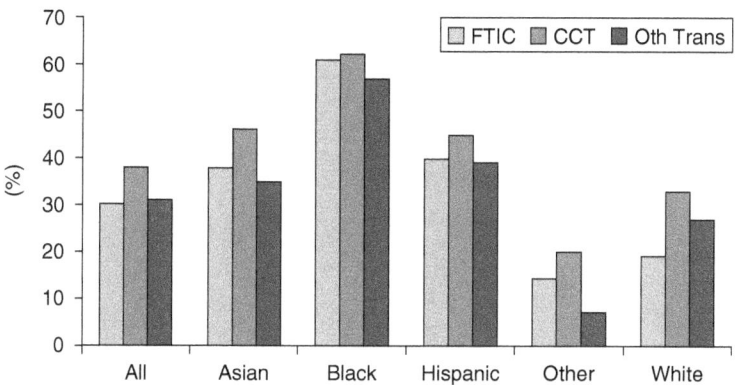

Figure 7.4 Percent Pell Grant Recipients by Enrollment and Ethnicity in Florida

Source: Adapted by the authors from the Florida Board of Governors Interactive University Database.

Transfer into Engineering

Using articulation data from Florida's Department of Education that lists declared majors for transfer students, a Chi square test reveals that differences in transfer rates into engineering programs are statistically significant across the five institutions (see figure 7.5).

For instance, UF had the largest number of students transferring into Engineering yet had one of the lower percentages of total transfer students. That is to say, a statistically significantly larger proportion of students who transfer to UF declare engineering as a major than is the case for students transferring to other institutions. Many of the students we spoke to at the feeder institution, Santa Fe Community College, were focused on the mastery of the community college pre-engineering major offered there. FAMU-FSU had the lowest percentage of students transferring into engineering, yet its combined numbers overall were among the largest numbers of transfer students. In contrast both FIU and USF each had equal percentages of students transferring into Engineering. As we have seen, more White students and male students are likely to transfer to UF, while more underrepresented minorities are likely to transfer to FAMU-FSU or FIU, mirroring the current populations of underrepresented minority students attending these institutions. These data underscore the assertion that White male students from more affluent backgrounds are more likely to enter engineering, thereby maintaining the status quo and assuring that engineering is a White male dominated field.

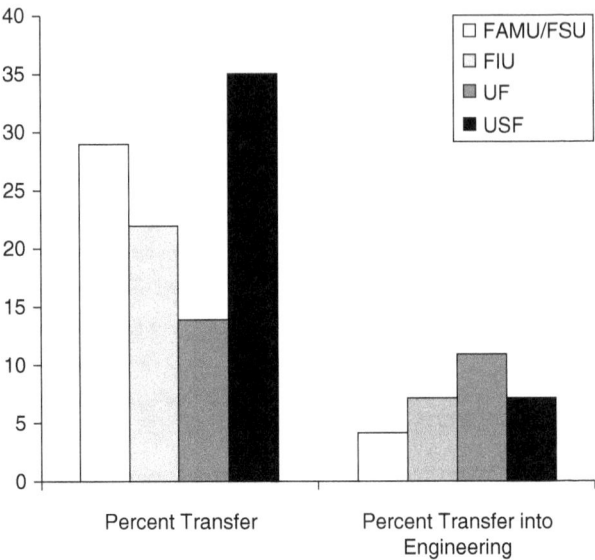

Figure 7.5 Percent Transfer and Engineering Transfer, 2006
Source: Adapted by the authors from the Florida Board of Governors Interactive University Database.

As is clear from this background, community college attendees are much more likely to be female, to come from less affluent families, to be over the age of 25 and also to be non-White (Black, Asian, Native American, or Hispanic) than those attending four-year institutions solely and are also underrepresented in engineering. Thus, understanding the transfer experience is critical if their pathways into engineering are to be improved.

The research presented in this volume grounds anthropological theory (political economy and practice theory) together with theory and principles from Industrial/Organizational Psychology (particularly with reference to culture and climate) in student experiences in engineering departments. This chapter applies these theories to an understanding of transfer pathways into engineering. Students from lower SES who attend community colleges may be at a disadvantage vis-à-vis more affluent students entering engineering programs. Prior research reveals a set of factors explaining why underrepresented students are less likely to transfer to four-year institutions than their White male peers. High school math coursework is an important predictor of STEM degree attainment in higher education (Tyson et al, 2008) and many underrepresented minorities come from lower SES. More affluent students are more likely to attend high schools with more resources and better science and math preparation in particular; therefore, these students enter universities with more social, cultural and human capital. Earlier studies (e.g., Seymour and Hewitt, 1997) show underrepresented minorities performing better academically in programs encouraging collaboration not competition; however, engineering is often a highly competitive field.

Similarly, women often report experiencing a "chilly climate" within engineering programs, that is, climates that are more male centered where females feel excluded or out of place because competition overrides cooperation in the work groups that are a common feature of engineering classes (Hall & Sandler 2002). As such, both underrepresented minority and female students may struggle to adjust to engineering programs. This adjustment is compounded by the difficulty of making a transition from a community college setting to a large, frequently impersonal university setting. We hold that many students may transfer into four-year institutions with inadequate skills or know-how, that is, the cultural capital, to easily make the transition into engineering. Examples of this cultural capital might take the form of knowledge about filling out a FAFSA for free student aid or the skills in navigating the bureaucracy of higher education. When students are uncomfortable or apprehensive about relying on institutional support, they turn to other students to get the information and support they need. While a reliance on social networks and other forms of social capital may ease the transition and provide underrepresented students guidance for traversing their programs, we argue that more could be done to make the transfer pathway more accessible and transparent to community college students.

Methods

To examine the transition experience of students embarking on the pathway from a two-year to a four-year institution, two different groups of students were interviewed using individual and focus group formats. Approximately 80 students representing four two-year community colleges across Florida participated in focus groups. Participants included students who were currently enrolled in community colleges, who were taking advanced math courses, and who were likely to enter STEM fields in four-year schools and students who had transferred and were currently engineering students. The students were selected from community colleges that have articulation agreements (acceptance of a full two years of credit) with four-year institutions in our study, that is, two-year schools that were feeder schools into the State University System (SUS). They included Miami Dade Community College (MDCC), Santa Fe Community College (SFCC), Tallahassee Community College (TCC), and Hillsborough Community College (HCC). In addition, fourteen students who transferred from community colleges were interviewed at the four engineering programs included in our sample. Finally, a workshop was conducted with students attending a Florida-Georgia Louis Stokes Alliance for Minority Participation (FLGSAMP) Exposition at the University of South Florida. Workshop participants described the pathway to graduation from their institution. This resulted in four visual representations of their perceptions of pathways to graduation which will be discussed as a part of a USF case study.

Community college students, faculty, and administrators participated in semistructured interviews and, in the case of students, also participated in focus group interviews. Participants were questioned on the following: (1) why students choose to attend a community college, (2) what challenges they experience and supports they received while there and, (3) student plans upon leaving the community college. We were not able to separate individual responses from the focus group interactions and, as a result, gender and race/ethnicity are not presented for community college focus group participants. In the SUS institutional interviews, researchers were interested in discerning similarities and differences between experiences at community colleges and four-year state universities. In addition, the experience of transferring from a community college to an SUS institution was of considerable interest. Using grounded theory, allowing themes to emerge from the data, we coded transcripts from the interviews looking for issues relating to community college students. We also collected background information (including income, ethnicity, residence, etc.) from all student research participants to illuminate demographic differences among community college students, transfer students, and students in the cohort of research participants who matriculated directly into a four-year institution (FTIC). The sample size does not permit comparisons based on sex or race/ethnicity and therefore student sex and ethnicity is presented only to allow the reader to contextualize findings.

As mentioned above, in addition to more traditional research methods, participatory workshops with University of South Florida minority engineering student clubs including The National Society of Black Engineers (NSBE) and students participating in a regional exposition sponsored by the Florida/Georgia Louis Stokes Alliance for Minority

Participation in STEM (FGLSAMP) underscored the differences in student perceptions of the steps necessary to successfully complete a STEM degree. These participatory workshops included an activity called 'Fishes and Boulders' where students visually outlined the steps, obstacles and resources needed to graduate with a bachelor's degree in a STEM field. Both community college and university students were given a bag containing colorful cut-out fish (representing steps or mini-goals); boulders (representing challenges in reaching goals); and seaweed (representing resources available in attaining goals). We then asked participants to draw a large stream on a sheet of poster paper and place the goal (graduating with a bachelor's degree in a STEM field at one end). Participants then had 30 minutes to use the cut-outs to "map" the pathway from starting in college to attaining a bachelor's degree. This activity proved most compelling when comparing the visual representations of obstacles and resources created by the two-year community college students and four-year university students.

The Community College Pathway

Community colleges in Florida serve a highly heterogeneous population of over 800,000 students. In our work with current and past community college students, four major factors led students to chose a community college over another type of educational experience: (1) cost, (2) an easier transition from high school (or for some students a "break" from school between high school and college to work or explore other opportunities) to a four year degree, (3) class size, especially as it relates to (4) a high level of personal attention from professors whose primary responsibility is teaching. In order to understand the experience of students transitioning from community college to a four-year university, we relied on current community college student's interviews and the reflections of current university students who previously attended community colleges. The SUS themes fall under two broad categories: (1) comfort level and support, and (2) strategies for success at four-year universities.

Findings from the community college students parallel the findings from transfer student interviews with regard to coursework and interaction with faculty and will be discussed together. Strategies for success reported by transfer students will be discussed further in the chapter. First, the two primary reasons students attend community colleges will be presented followed by the presentation of results concerning coursework and interactions with faculty.

Why Students Choose to Attend Community Colleges

The first factor students identify is cost. One student from Hillsborough Community College (HCC) characterizes the low cost of community college compared to a local university this way:

> I [went] to USF asking for admission and those guys told me I could pay out-of-state fees and they told me it would be... about $16,000 a year. Then somebody told me to come check out HCC. They offer the same classes over here. So I came over here and they told me out-of-state fees per year would be around... $6000 if you take full-time.

Other students echo this sentiment adding that community colleges are also more cost-effective if you are undecided about a major or if you aren't sure college is "for you." These students feel that the lower cost allows them to try more new things without risking a great financial loss.

Many students express apprehension about the transition from high school to higher education and believe the community college help ameliorate those fears. Two focus group participants summarize these fears:

> I1: "A lot of kids get freaked out by going from high school onto a college as big as UF and your GPA drops really bad."
>
> I2: "One of my friend's GPA wasn't as strong as it was in high school even though he knew all of the material. Basically it was too overwhelming." (Santa Fe College Focus Group)

These sentiments are echoed by other students who also rely on preexisting social networks to help them succeed in college. Several students at each school indicated that they choose to attend a community college because their friends planned to attend the same school. One Latino male student entered the Navy immediately after high school and was starting higher education now that his military service had concluded. He chose to attend the local community college because a Navy friend was also starting, and he did not have the grades for a four year college acceptance. The advantages of having a friend at the community college outweighed the prospect of being alone at a four-year school. Social networks play a role in determining where a student will begin an academic career. Parents also figure importantly in determining where their children go to school. One Latino student at Miami Dade College indicates her father did not want her to leave Miami:

> [My father] thought it would be a good thing to [start at Miami-Dade Community College], so we agreed to come here. I actually wanted to go out of state, but he doesn't want me to go. And also there are other reasons,

activities that I'm involved with. It's just...it makes more sense for me to be here right now.

Another student notes his parents attended a community college and shared their positive experiences with him. While a different student spoke about her parents' negative four year college experience with large courses. Her parents encouraged her to start at a community college to get a more personalized experience. In sum, parents, friends and other people in a student's social network provide guidance and support and determine where students attend post-secondary school. The influence of a student's social network is well documented in the literature (Perna & Titus, 2005).

Class Size and Faculty Support

The third and fourth factors, class size and faculty support, affect community college students' decisions and parallel findings for transfer students at universities. Regardless of whether the students were pre- or post-transfer, the majority of all students interviewed indicated a small class size and faculty whose primary responsibility is teaching are major factors in deciding to attend a community college. Students points to the increased level of personal attention that small classes and dedicated teaching faculty give them. Research suggests that community college students are better supported by faculty at community colleges, i.e more individual attention, smaller classes, but that it may make the transition to four-year institutions harder for the students (Caporrimo, 2008). This research supports the contention that the sense of support and community at a two-year institution may indeed make the transfer experience more difficult as we found that students often compare the benefits of community colleges to their four-year institution. We found the primary difference between the community and four-year institution could be described as the students feeling more comfortable at community colleges: they are more comfortable in classes, in seeking help, and most salient to them is their inter-personal relationship with professors.

Class size. All but one of the transfer students individually interviewed are more comfortable in smaller classes. One even favors community college classes overall (UF Hispanic male student). Another notes that while large classes are helpful because there are more students to get to know, it is hard to ask questions (USF White female student). Still another believes community college courses are better for preparation—especially with calculus since these classes are smaller allowing for more individual student participation (UF White male student). Another student sits in front, but complains about not hearing professors, noting the most engaging teaching in large classes depends on the professor's approach (UF Asian male student). The balance between the small size of community college classes and the expertise of research-oriented professors in four year institutions is a delicate one. Accessibility in and out of the classroom seems to be very important for student success. One student states clearly: "smaller groups are usually better than big ones and there's a couple [of] classes in my department that get so full, so many students you cannot really appreciate the class or focus on it" (FIU Hispanic male student). Being known by the professor is a product of small classes and is conducive to comfort (FAMU Black female student). Several students suggest that a class size of 15 is an ideal class size;

> I mean the stuff we're learning isn't the easiest stuff in the world so sometimes when you raise your hand the teacher answers the question in a real terse way but sometimes that's not enough. Sometimes you need to say one plus one plus one is three...sometimes it's not the most obvious thing in the world. (FIU self-identified "Other" male student)

Engineering is a rigorous discipline, the students find, is better taught in smaller classrooms where the students were afforded more personal interactions with faculty and are more comfortable asking questions.

Half of the students who participate in a focus group said the classes at their universities are much harder than in the community colleges. In no instance do students say that they believe community college classes to be harder but there are instances of interviewees describing how the community college classes prepared them by serving as an incremental transition between high school and university coursework. Two USF students say their university professors require homework to be handed in which was different from their respective community colleges. They joke that classes are empty at USF the day homework is due because everyone is out finishing it (USF Black male student). One community college transfer student at UF feels he had to spend a lot of time "catching up" with the other students and the coursework itself (Asian male student). He also thinks community college transfers should be interviewed before entering universities because not all of them can handle it; he thinkst they should put effort into sorting out the "bad apples." This discourse is indicative of lingering stereotypes of community college students as less capable than university students and also underscores the lack of consensus and understanding of what university faculty expect from students.

Faculty availability and support. Community college students' views of faculty contrast with perceptions of professors at four-year institutions.

Faculty are accessible in community colleges; they are inaccessible and unavailable in universities. Students describe community college professors as being "always available" outside of class to answer questions and indicate that this was a major resource:

> Nine times out of ten they're in their office and you say... "I don't understand this problem at all"... and they ask you questions and they walk you through it to see if you get it on your own or, if it's difficult, they [say] this is exactly what you need to know. (Miami-Dade College student)

Community college students express a desire to belong in a community with other students and with the instructor. They express frustration at the idea of simply being an anonymous student in a class. One student characterized this sentiment thus:

> The thing at UF is that it's very easy to get lost in the anonymity of just being a number and you can use that to your advantage or...some people lose a sense of their self. I came from a very small high school where I knew everyone and I knew all my teachers and I went to school with the same kids from Kindergarten through 12th Grade. And going there was a bit of a culture shock just being a number and not knowing my teachers and my teachers not knowing me. Just showing up and having to bring an I.D. to prove that I am who I am so I can take my exam. (Santa Fe College student)

The research pursuits of engineering faculty are also a common theme in the transfer student interviews. Professors whose main focus is research have little time or interest in students and their office hours are always backed up (FIU Hispanic female student). Conversely, another student believes the instructors at his community college put their time into teaching and not other pursuits. The community college instructors are committed to teaching while his university professors are concerned more with research and, "to get money to the [school]. That's their main goal; that's what gets them promoted among the ranks" (UF Hispanic male student). Similarly having capable teachers is emphasized: "One thing is to know your topic. The other thing is trying to teach it so all students can understand it" (FIU Hispanic male student). Transfer students from community colleges recognize the importance of research for university professors compared with their community college counterparts. A focus on research is seen as a distraction to transfer students who appreciate their community college professors' commitment to providing quality teaching.

Another salient difference between community colleges and four-year institutions, expressed by three transfer students (White female students at USF and UF and a Hispanic male student at UF), is that instructors in community colleges focus on application rather than theory, leaving one of the female students feeling unsure what she should know in her classes. Another student is afraid of being behind: "At first it was crazy, I was real nervous about things" (UF Black female student). An Asian male at UF believes that, "The only way you can learn application here is by doing the research and putting effort into it and working with some professors." One student believes there should be more labs to learn application of theory. (FIU Hispanic male student). Another expresses insecurity about what other students may have learned in their university classes that she did not acquire at the community college (USF White female student). The focus on application in community colleges classes provides the transfer students with real life skills; however, these students believe they were not prepared for the focus on theory they now experience in their university engineering classes. Their desire for application is echoed by university students who are not transfer students. To overcome the sense of missing theoretical knowledge or just being behind the other students, transfer students relied heavily on one another for help.

One third of transfer students we spoke with expresses apprehension about approaching professors for help, believing that professors are unwilling to help, or do not have time. Specifically, they worriy their questions are not appropriately sophisticated for professors to answer thus making them look unprepared or 'stupid' in the eyes of others; one student thinks questions have to be "worthy" to present to a professor (UF White male student); another is afraid to ask what he thinks may be considered as "dumb question, "There's some you can tell on their faces that they probably got better things to do...maybe it's a dumb question for them but I don't believe that there is such a thing as a stupid question" (UF Hispanic male student). He also thinks the professors are "impatient" when asked questions outside class. Seeking help from professors outside class time is an issue for a number of students. One student said only half the students in his class interact with professors outside of class and that it is partially the students' fault and partially the fault of professors as to why more students do not seek help (FIU Hispanic male student). We see here the students' use of words like, "dumb," "stupid" and "impatient," clearly demonstrating their belief that they are somehow not worthy to speak with faculty at the four year institutions they attend. This reluctance on the part of university students to engage professors contrasts sharply with the comfort found previously attending community colleges where, owing in part to the small class size, they were able to get to know instructors. The small class sizes that are common at community colleges enable students and faculty members the opportunity to get to know each other on a personal level.

Students mention that community college instructors provide emotional support that encourages them to succeed. Students pointed to professors knowing their names and saying hello in the hallway. These gestures, while small, are highly meaningful and indicative of the welcoming and nurturing culture and climate of community colleges for all students interviewed. Most students believed smaller classes and higher levels of personal attention held them more accountable, gave them assistance and reassurance and helped them to be more successful academically. Again, paralleling the findings from the community college students, transfer students whom we interviewed speak of their community college professors more personally than their university professors. One student's description of his community college teachers centers on emotional support; his community college instructors are more apt to offer words of encouragement while his university professors were concerned with their intellectual talent and acumen: "University professors were inspirational because they were "geniuses"- a lot of the professors you look at and you can't help but be inspired...it's hard not to be inspired when you just hear the professors do what they do sometimes" (UF White male student). Hispanic male students at UF believe "good" professors have prior experience in industry principally because they had to learn interpersonal skills there. A "good" professor is someone who is able to develop a personal relationship with his or her students and who challenges students while in class, a professor who is, "being aggressive with the content and trying to bring out everything that [the student] has got to offer" (UF White male student). When asked what could be improved in his program, another student remarkes that faculty should be more encouraging and personable; only a few, he notes, can remember student names (FIU Hispanic male student). A professor's personality is also important because it can make students feel more comfortable. While this student does not compare her previous university professors with those she has in class currently, she does discuss the importance of support and mentions one of her professors who, according to her, keeps saying, " '[t]his is not easy where you choose to be. And if you choose to be here, you have to care about it.' So they encourage you" (FAMU Black female student).

Beyond personal attention, community college students point to classes taught by faculty, rather than graduate students. At Santa-Fe Community College, the feeder institution to the University of Florida, one focus group participant remarks, "You don't get passed around graduate teaching assistants like at UF. Here you can actually go ask your teacher; he'll be able to answer your question instead of going through three different TA's that have no idea what you're talking about." Many community college students interviewed appreciate this more direct approach to education; students interact solely with the professor. Most of the community college students interviewed indicated that they do not take advantage of community college resources such as tutoring centers or homework help desks but instead prefer to work independently and seek help directly from a professor when needed. When community college students transfer to four-year schools, and no longer have direct access to professors, the question remains where they will seek out help. In addition, this finding has implications for the allocation of resources and infrastructure for both community colleges and university engineering programs.

In contrast to the findings described previously, indicating students are more comfortable in smaller classes and receiving personal attention accorded by community college instructors, two male students at FIU, both presidents of student engineering societies, feel comfortable approaching professors. Their comfort with professors may be related to their involvement in student clubs as faculty members are routinely drawn into club functions. Students who are highly involved in the student organizations such as SHPE and NSBE—and with the organization's faculty mentor- seem better able to overcome apprehension about engaging faculty. In short, transfer students gain, by way of their organizations, skills in approaching faculty (cultural capital), as well as the networking skills to facilitate this (social capital).

At a four year university, effort must be made to facilitate the process of transferring from a community college. A Hispanic male student at UF experienced a particularly difficult transition from a community college to a four-year university, despite a mentoring program for new transfers by older students. His grades dropped because the classes were exceptionally difficult and he believes he was not given guidance about course load and sequencing. He believes his academic program at UF had potential but "disintegrated" as soon as the semester began- as he was unable to keep up with coursework and began failing classes. Getting to know one's way around campus was difficult for two female engineering students with whom we spoke. Orientations that focused on the campus geography would benefit transfer students. In addition, an adjustment period for community college transfer students might be considered and support offered as students learn to modify and adapt their community college cultural capital to the cultural capital of their four-year institution. This might take the form of a for-credit class or optional co-curricular opportunity for transfer students that would focus on creating new linkages and helping to establish support groups.

The Pathway: Transitioning from AA to BS Contexts

Strategies for Success after Transferring

According to transfer students, participating in group work and developing social networks are the primary strategies for success in an engineering program at a university. Specifically, transfer students discuss reaching out to other students as mentors and advisors and connecting with students already in the program. These findings echo the sentiments of students throughout this research: when students do not come to college with the cultural capital of how to navigate institutional support, they rely on other students for emotional support and for skills required to succeed in the program.

The importance of social networks. As we saw in the previous section, students are apprehensive about approaching professors and several report negative experiences with professors outside of class. One White male student believes professors dismiss students coming to them for help by sending them to teaching assistants. A White female and male from USF and two Black males from USF discuss the topic of tutors or graduate/teaching assistants. The White female student had met with teaching assistants at her community college, although she did not seek them out at her university. The two Black male students believe the tutoring center is a great place to meet other students to study with, but are ambivalent about the efficacy of the tutors themselves. Another student is more comfortable seeking out teaching assistants rather than seeking a professor's assistance since, "they're usually closer in age to us so they kind of understand if you have a problem" (FAMU Hispanic female student). Another does not feel supported by the department, "My support I would say and I really don't have any from the department...my support is really coming from the people, the students that I've met" (FAMU Black female student).

Student organizations are, as we have noted, important in providing support for transfer students, although some students were initially reticent to join for fear of overcommitment. Others describe the importance of student organizations in providing academic support for students, offering emotional support and encouragement, and helping make the transition from a community college easier by offering supportive networks and guidance about the program, "Yeah, there's a lot of organizations that allow students to kind of come together and I think that in a major such as this one...you can't do it by yourself" (UF White male student). The importance of organizations for minority students, in particular, is shown by their involvement in activities in engineering societies. One student reports it took him a whole semester before he made friends, and that because of this ethnicity as a foreign nation he was not able to meet anyone. Another student, now actively involved, believes that student organizations are important for students to meet people in the event they need help in the future (Hispanic male student). Another student advocates the importance of social organizations, academically as well as socially, believing that groups can help students feel they fit into an engineering department (UF White male student). One Hispanic student from Latin American chooses to seek clubs outside of engineering; he finds camaraderie with other Hispanic students by playing soccer.

When students were asked about their level of involvement, several refer to the amount of engineering homework they do as a defining factor of their involvement, more so than participation in a club. One student claims that after she transferred to her current university, she spent all her free time going back to visit her old community college. She now has friends at her current institution, but prefers to study alone, "they don't seem like they're really interested in trying to meet people. They usually are trying to get their work done." However, she considers herself involved—considering, "I'm involved in it to the point that I eat, sleep, breathe Engineering." She concludes, however, by stating that she would like to get more involved (FAMU Black female student). Another student also preferred to study alone and did not seek out other students initially,

> I think it has to do with that I feel like if I'm in a large group of, you know, men, that they're the majority. Like I have some kind of pressure that maybe they won't believe what I'm saying... So it's like you know I'm like searching my pathway to go. (FIU Hispanic female student)

The two students from this interview sample who preferred to study alone were females who say male students make them uncomfortable. Some projects supported by the National Science Foundation STEP Type I awards advocate the importance of "birds of a feather" study group composition.

The importance of group work. Half the transfer students interviewed indicate they are more comfortable going to a friend or classmate before going to a professor. All participants mention the importance of group work, although the reasons for relying heavily on group activities vary. All believe it essential to traversing the program successfully. White and Hispanic male students at UF appreciate that students find encouragement from groups; they in fact mention their friends push them to do better or encourage them when they need it. Another White male student believes group work prepared him for being a professional engineer as "they work in groups in industry." Still another, an Asian male attending UF who initially worked alone, sees that groups are a way to double check if your work is correct. A student who was at first apprehensive to work with others because she did not know any of the students assigned to her study group, learned to approach other students who were studying:

> I don't know these people, I don't want to intrude on them but then, you know,...I came here a couple other times and I'd see people by themselves and I went up to them...and they actually had a lot more questions than I did so that was good to know...so that's pretty much how I get to study with people since I don't really know anybody on a personal level yet. (USF White female student)

Groups and friendships also offer emotional support (a benefit of being in many types of social networks). Two Black male students in a FAMU focus group believe that the community of FAMU, "It's kind of like a family or a brotherhood almost." Peer/student support created via group work is very important:

> Working together is a matter of getting through, you know? And sometimes you're down and they push, sometimes you know they're down and you push them and it's...like...you got to do the whole team work, the buddy system...like in the Marine Corps. (UF Hispanic male student)

Another salient theme is requesting guidance from students who have been at the university for some time. A White female at USF states, "I mean now that I'm already a junior and other people here have been going here since their freshman year they kinda know everything so I can just talk to them." She introduced herself to other students and was relieved to hear that they also had questions from class; she believes they can help her adjust and "point her in the right direction." She thought the best characteristic about her university is other students' willingness to help her and not blow her off.

Student organizations such as NSBE are a particularly important venue for meeting other students, forming networks, and receiving guidance about the department. One student wishes he had come to his university while still enrolled in his community college to speak with students about their experiences there—most importantly about how to sequence courses and avoid becoming overwhelmed by the course load. As he had previously served in the military and thus his use of a service metaphor to describe other students is particularly apt:

> I had to make up a strategy on my own and the only way I was able to do that was by gathering intelligence like we do in the military and the best intelligence you can possibly get is by talking to people that are out there in the field...your fellow students. (UF Hispanic male student)

Another student does not have a faculty mentor, rather he found a PhD student as his mentor (UF Asian male student). Many research participants, in lieu of seeking out a faculty mentor, rely on peer networks, not surprising given the importance of peer networks uncovered in previous research. In particular, student peer to peer tutoring and group work builds a set of desirable skills from the future engineering employer's perspective including the capacity to collaborate, work as a team member and enhance the overall quality of students' learning.

FGLSAMP Expo Findings

Researchers conducted several participatory workshops with students who are active in the Florida-Georgia Louis Stokes Alliance for Minority Participation (FGLSAMP) at USF. We conducted these workshops with students active with FGLSAMP at USF and those students who participated in an FGLSAMP Exposition. Many Expo participants were high school and community college students. As previously mentioned, the goal of these workshops is to discern the obstacles student perceive to successful graduation.

There are striking differences in what students attending four-year institutions perceived as obstacles compared with those still attending high school or enrolled in community colleges. For example, on some of the larger "boulders," community college Expo participants incorporate critical life circumstances: pregnancy, eviction, DUI, sickness and drugs. They drew a fisherman on the side of the river who actively "fished" students from the stream toward the lake representing the goal of graduation. To these students, the boulders represent challenges, and also include particular individuals whom the students believe worked against their academic success. Students were clearly aware life circumstances might become overwhelming and take them off the path to graduation and they needed to be on the lookout for people who did not have their best interests at heart. They list more financial resources as well including money, clothes, food, a reliable vehicle, resale shops and a bus (representing transportation).

In contrast, the USF students had much clearer perceptions of the specific steps needed to graduate with degrees in engineering. They list internships, time management skills and going to class as specific actions getting them to graduation. Their obstacles are not financial but rather about getting distracted by partying, spending time with friends and romantic partners and getting burned out from studying. This is not to say that the four-year students are unaware of or not impacted by financial constraints but that these burdens do not loom as large as social distractions as indicated by both the relative size of the "boulders" selected and by the number of times it is mentioned on different posters. What we found was that the students who attended the Expo were far more concerned with financial burdens and life circumstances than were the four-year engineering students. Furthermore, university students had clearer perceptions of what is required to be a successful STEM graduate.

Conclusions and Policy Recommendations

Based on our analysis of the emergent themes we recommend several practices and policies that are most conducive to a successful transition from a community college to a STEM field at a four-year institution. We also identify which practices and policies are most problematic for student success. Students do not "leak" from a pipeline- many are not able to learn success strategies or are faced with financial obstacles they cannot overcome. Community college students are increasingly heterogenous and upward of 40% of engineering majors have attended community colleges at one point in their academic career. As such, both obstacles and strategies for success must be examined.

Class Size

Students interviewed express dissatisfaction with large classes and prefer the more individualized attention found at community colleges. Several students attend community colleges for this very reason. They report feeling more comfortable asking questions and like the feeling of being recognized as an individual, in contrast to feeling lost in the crowd. The same holds true for non-transfer students enrolled in four-year institutions (see chapter four). This echoes the contention that smaller class sizes are an effective way to improve student performance (c.f. Nye, Hedges & Konstantopolous, 2000). We recommend that policy makers and university administrators more readily consider empirical data in support of smaller classes to guide the drafting of education policy.

Faculty Availability and Support

When taken as a whole, our data supports previous research that small classrooms and encouragement provided by faculty to students are strengths. Given the students' apprehension or uncertainty about faculty office hours, one recommendation could be that professors require all students to come to their office hours at least once in the first week of school. We realize the time commitment this would entail, and perhaps faculty could place sign-in sheets outside the door to determine that the students were at least able to find their office. The strategies would help students overcome the initial fear of approaching faculty and would particularly benefit transfer students who may not know where faculty offices are located.

Much of the previous literature on the community college pathway calls for improved communication between two and four-year institutions, particularly for the improvement of advising. For example, Flaga suggests that "transfer shock" can be mediated by more structured advising relationships between them (2006). While we do not argue with this assertion, our data suggests that students are ambivalent about formal advising, that is they may be unsure about the process or they may have experienced poor quality advising or at the least heard rumors about it. While rumors are easy to dismiss from an academic standpoint, we found that students rely heavily on other students for information about how to get through engineering programs. Also, while our research found that several of the institutions visited had transfer programs or orientations only one of the transfer students mentioned these programs. While this could reflect sample bias it could also indicate that students did not utilize these programs or that they had little lasting effect. Again, our data suggest that students rely heavily on informal networks for their information such as other students and engineering clubs.

The community college students suggest online advising where they would be able to read about classes they had to take and program requirements. We are aware that these resources exist at the institutions that participated in this study, but the interview data reveals that students are not utilizing them. Whether the online tools are not available or hard to use was beyond the scope of this research but we hold that it is an important topic for future research. Problems with advising and program requirements were a recurrent theme throughout the larger project and held true for the community college and transfer students as well.

Importance of Social Networks and Group Work

The transfer students interviewed report seeking out fellow students as the means to "gather intelligence" about the program. Furthermore, student organizations featured heavily in the discussions, though several of the students express trepidation about joining immediately after transferring for fear of overextension. Our recommendation is therefore related to student mentoring to supplement formal advising. First, we recommend that upper-level students who have been at the university and who are familiar with the departmental policies and practices serve as student advisors/mentors to transfer students. We suggest former transfer students as they may best understand the transfer process and the experience of transferring. We should note, though, that students serving as student mentors should receive training about departmental practices and policies so they can provide accurate information. Second, all the students who are involved with clubs comment on how beneficial the clubs are to their undergraduate experience and note that student help is essential for getting through the program. We recommend that transfer students be given information about social and academic clubs and that club members reach out to transfer students, or clubs targeting transfer students be created. Flaga (2006) recommends that advisors from four year institutions go to community colleges to meet with students before they transfer. We would argue the same for student outreach and liaising. Finally, we recommend that transfer programs already in place be bolstered so that they do not, "disintegrate" as a Hispanic male recounted happening at his school.

Non-school Obligations and Financial Constraints

The workshop conducted with the FGLSAMP participants who were four year students described their challenges to graduation through the Fishes and Boulders workshop. While the students at USF report specific obstacles such as advising and class availability, many of the community college participants report obstacles different from those of their four year counterparts. Besides reporting college partying to be a major obstacle, the obstacles facing the four-year students are primarily process oriented: they have to do with filling out forms, or scheduling appointments and classes. In contrast, the obstacles facing community college students are structural and the result of political economy. Financial constraints and life circumstances such as pregnancy and transportation were seen as much larger obstacles than the practical or social obstacles reported by four-year students. Social class differences between community college students and university students are rooted in the real practicalities of structural factors such as poverty and lack of public transportation. Furthermore, seemingly personal obstacles such as unplanned pregnancy and substance abuse are linked to socio economic position. While we are not asking for transfer students to have different academic standards, we hold that many community college students face structural barriers and challenges not experienced by more affluent students. As such, the difficulty in transferring should not be "blamed" on the student but rather understood in the nexus of social inequality.

Addressing Student Uncertainty

We hold that attention should be paid to psycho-social effects of transfer and adjustment. Many transfer students report trepidation approaching university faculty and uncertainty about how universities function. This uncertainty leads students to avoid professors' posted office hours, avoid joining social organizations, and avoid utilizing campus resources. Students rely on one another for advising and class information and while we recognize the benefits of social support and networking, we argue that transfer students would benefit immensely, gaining important skills and cultural capital by learning how to access faculty and campus resources. We recommend trained student mentors would benefit transfer students and would address the need for social and cultural capital necessary to obtain a degree in engineering.

> *So it's like, you know, I'm searching my pathway to go...*
> —Florida International University Hispanic female engineering student

We challenge educators and researchers to broaden their perspectives and policy recommendations to account for variation in student experiences. Community college students may be at a disadvantage because of their social class positions; they may have to work while going to school and may have to support family members or children. These factors can constitute obstacles to their success. This does not negate their agency, however, and many are able to successfully transfer to four-year institutions and graduate. The decision to attend a community college is generally highly strategic, in our research we found it is used to either temper the transition from high school to university or to gain access to more personalized attention in smaller classes. For successful transitions and improved program efficacy, institutions should consider focusing on programs to help transfer students learn additional strategies that other students may already have: how to utilize campus resources and how to bridge the gap between faculty and student expectations. Recent research underscores the importance of nurturing students' sense of belonging, a virtue of attending the much smaller community college, "Given the importance of sense of belonging for promoting student persistence and academic achievement, colleges and universities should find ways to facilitate these interactions with diverse peers that lead to positive educational outcomes" (Locks, Hurtado, Bowman, & Oseguero 2008, p.280). This is to say when students come with little cultural capital about what is required to succeed as an engineer, university policies and practices must aim at building cultural and social capital. In conclusion, a broader understanding of student experiences along the engineering pathway will allow for more students' needs to be met and for more students to remain on the path to graduation. It has been the aim of this chapter to provide the perspective of the community college experience for all students, but especially those students underrepresented in engineering.

Chapter Eight
Voices from the Field: Strategies for Enhancing Engineering Programs

Kathryn M. Borman, Will Tyson, and Cassandra Workman Whaler

Introduction

In our concluding chapter, the aim is to make concrete recommendations building from the analyses presented in chapters three to seven in this volume. The goal is to assist undergraduate engineering programs, particularly those located in public universities, in strengthening their departments. The research informing this volume was a multi-disciplinary, mixed-method study which combined qualitative interviews, observations and focus group interviews with quantitative faculty and student surveys. This volume presents primarily qualitative data as interviews with students, faculty, administrators, staff members including counselors and advisors reveal strategies for enhancing undergraduate student experiences as well as revealing reasons why students left engineering. Faculty, administrators, and staff provide testimony based on their experiences with, in some cases, generations of students whose actions may lead to switching from engineering, poor academic performance, and/or delays in degree attainment.

These interviews suggest how to develop a more inviting culture and climate for engineering programs in large public universities by improving how these programs might be structured to enhance student learning.[1] To persist in engineering, students must be prepared to re-take difficult classes, work cooperatively and competitively with others in their programs, and fruitfully engage in internships and mentoring arrangements that are pathways to jobs in industry. These actions are stressed by faculty, administrators, and key staff who discussed what they see constituting successful pathways to degree completion. Other strategies for student success include building an enduring interest in engineering, addressing insufficient preparation in mathematics, and persisting despite setbacks and problems. Students also expect institutional support and create strategies for persistence in response to what they see as a lack of support from faculty, administrators, and staff.

As we have seen throughout the book, certain students, because of their position in society or because of how they are socialized in our culture, are at a disadvantage vis-à-vis White males to succeed in engineering. This is not to say they are unable, rather, we hold that because of their relative positions, it is critical to understand their experiences as undergraduates in order to foster their identity as an engineer. We follow researchers such as Seymour and Hewitt (2007) and believe that engineering education should be improved for all students; yet the unique experiences of women and minority students are critical to uncover.

Building an Enduring Interest in Engineering

Chapter four shows that early access to career information is essential in attracting more women and minorities to STEM careers, particularly to engineering. Research, including this study, demonstrates that women and minorities are more likely than White males to become interested in STEM fields through participation in specific advanced and targeted programs during high school (Margolis & Fisher, 2002; Moore, 2006). This reflects the cultural tendency in our society for teachers, parents, and others to encourage males more than females to enroll in advanced science courses. Many more men in our sample reported becoming interested in engineering before high school than did women. Men became interested as boys through play activities that model the work of an engineer such as creating structures with Lego building blocks. Women became interested through specific programs targeting girls and because they had family members who were engineers. One woman describes her initial interest in engineering:

> Oh that's an easy one. It was the summer right before I was going into 8th grade and my parents enrolled me in a camp back at home. It was an Engineering/Mathematics/Science camp....I got interested in engineering...it was like just the things that we were doing in general really interested me about Engineering. And then once I like tried it...I wanted to be an engineer. (FAMU Black female student)

This young woman's father worked at the university hosting the program and knew enrolling his daughter would be beneficial to her academic development. Mentors and role models are more important to young women in encouraging their early interest and participation in the sciences than is the case for young men (Frestedt, 1995). Another young woman described both a significant activity and assistance from her father, an engineer, in completing that activity:

> Well, it was in middle school in sixth grade because we had to build toothpick bridges for the toothpick bridge contest and it was required that year... and my dad helped us because he's in the engineering department and we actually both really enjoyed it, and we ended up getting third place at the competition... of all the nearby schools. So the next two years... I did it by myself and I ended up getting second and then first. So that was when I knew. (USF White female student)

Interestingly, two White female students reported becoming interested in engineering though toothpick construction competitions in high school. As engineering is not generally offered in a high school curriculum, the two young women were able to gain an understanding of engineering as a discipline through their relationships with a close family member. A Hispanic female student sums this up, "My father was an engineer so I had an idea of what engineers did. Not many of the other people did know" (FIU Hispanic female student).

Several minority male students we spoke with mentioned taking advanced math and science coursework while in high school that led them to engineering. As discussed throughout this volume, certain minority groups are underrepresented in engineering: African Americans, Hispanics and Native Americans. The reason for this is complicated but one driving factor is that many of these minority students may come from lower socio-economic strata and may therefore have been more likely to attend lower performing schools. Similarly, many underrepresented students may be the first of their family to attend college which may make an already difficult experience all the more difficult without the cultural capital, or know-how, or guidance to succeed in a program. Unfortunately, as there are often few minority faculty in engineering programs, minority students are also less likely to find mentors or role models within engineering programs.

As such, engineering programs must work with students to nurture genuine interest in pursuing an engineering career especially among women and minority students who are likely to become discouraged because there are simply so few of them. Chapter three shows that many switchers did not have an accurate idea of what to expect from either an engineering career or an undergraduate major in engineering. Engineering programs should contrive to provide more activities in local communities including sponsoring programs similar to toothpick construction to nurture an interest in engineering among middle and high school students. Colleges and departments of engineering may also want to coordinate these programs with community colleges to expand their reach beyond large and mid-sized urban areas where most universities are located.

Bridging Gaps in Academic Preparation

As shown in chapter three, many women and minority switchers did not have the previous course-taking and achievement in mathematics in preparation for an engineering curriculum, primarily because they did not know how important high school mathematics preparation would be in their pursuit of an engineering degree. Several switchers report not knowing what to expect from the university and the undergraduate engineering program specifically:

> I'm not a strong math student, and these are things that my [high school] counselor knew, but I don't really feel like she took the initiative to try to show me places where I can go and try to get that developed before I got to college. So when I did get to college it was very overwhelming for me. I felt like I was very under prepared in the math section. I wasn't prepared for college math. (USF Black female switcher)

Despite faculty concern with lack of rigorous student preparation, there is little that departments can do to prevent underprepared students from entering an engineering program because in most public university programs, undergraduate admission is monitored by the admissions office not the department. Nonetheless, strong math and science skills are critical to the successful completion of undergraduate engineering degrees. Involving engineering programs early on in undergraduate education and before students' junior year is critically important. As this FIU administrator explains, underprepared students and the faculty who teach them are stuck with each other:

> The biggest difficulty we have is that math component. . . . They're under prepared in the math area. So you know we have to work at getting them there. And we can't say, "Oh if you can't start in Calculus, you can't be an engineer." . . . .While some people would like us to do that, I can't in any circumstance sit down and say, "Because you're starting in College Algebra, you have to do College Algebra, Trigonometry and then Calculus I, sorry you just can't be an engineer." (FIU Administrator)

Engineering programs must work with students to address deficiencies in their preparation while nurturing a sense of genuine interest and commitment to engaging in rigorous coursework, a quandary for engineering programs that we investigated in this research. Faculty and administrators realize that not all students are prepared for engineering, programs but cannot deny admission to all students based on their lack of academic preparation. The FIU administrator continues:

> You know…so we try to work with the students and try to work with the departments to help them understand that. You know the students that are coming in may not be following the exact…plan that you have because they're not there. They're not strong enough in calculus. (FIU Administrator)

Thus, some departments coordinate pre-college programs to help students get "there" and to bridge the gap between their lack of academic preparation and the expectations faculty hold for their achievement. A UF administrative coordinator described the purpose of such programs:

> We have the highest percentage of National Merit Scholars on this campus in this one [engineering] department. Okay, but then at the same time I have students that are coming into this department who didn't get calculus in high school. So I have a very big broad range of kids who are interested in engineering and are very creative and want to be problem solvers and that's why they're interested in engineering. They like math and science, but their preparation was night and day. And so we have the ability to take a student who comes in from not having had all the resources in high school and not having had all the preparatory classes in high school and we're able to help that student transition in to the University of Florida and also navigate the courses in such a way that when they leave here in four years you won't tell the difference. (UF Administrator)

This administrator has confidence in the institution's capacity to provide guidance and support to bring those students lacking adequate preparation up to par with their classmates and not blame high schools for lack of access to rigorous coursework. In fact, UF has an exemplary and well subscribed program, STEP UP, run by a very popular administrator. Our survey results reinforce this point, showing that women and minority students appreciate pre-college transition support programs and find them extremely helpful. Although these programs are essential in encouraging retention among women and underrepresented minority students, they are only one solution to the problem described by administrators. Another solution is for engineering programs to have more input into which students are admitted into engineering as we have mentioned previously.

> We have no say in admissions. We just take what they give us and that's been one of the concerns that we have is if we had a little more say in admissions. Now the counter argument is "look, we're taking the top four percent

of students in Florida, you wouldn't be taking students out of this pool anyway" and maybe they're right, I don't know. (UF administrator)

UF Engineering has the luxury of picking from the best and brightest students in Florida; however, this administrator would appreciate a more active role in evaluating student profiles to determine who might have the best chance at completing an engineering degree. This administrator continues:

> But sometimes you see things in students' activities that tell you clearly they're going to be great in engineering because they're very applied, they're good at making things, maybe they didn't do so great in their high school calculus class, but they're great outside of the classroom. They are probably some of those that we would select, maybe that they wouldn't, but absent any, like I said, we basically take what they give us. They just show up and then we try to do the best with them. And I bet a lot of students we get that say they want to do engineering probably really don't. I think if we were to talk to them during admissions, we could tell whether or not they're really into it. So our attrition rate is horrible. We only give bachelor's degrees in engineering to maybe 50% of the students who come in as freshman saying they want to be engineers. (UF administrator)

Transition pre-college programs and greater programmatic influence in admissions into engineering represent two structural improvements that could advance program efficacy among engineering departments. Using a mathematical analogy, pre-college support programs increase the numerator while more influence in admissions increases the denominator. Better identifying students who really want to be in engineering and have both the academic capital and genuine interest to enter engineering as freshmen allows administrators to prepare programs easing students' transition into the program. This option also allows other students to delay their entry into engineering until their junior year. One UF administrator describes benefits of this program:

> We now have the opportunity for an incoming freshman to become an electrical and computer engineer. They can declare themselves as...[an engineering major] right off the bat or they can be in the college of liberal arts and sciences at which point they can make a degree determination at some point, at the junior level or anywhere in between....It gets them advising in our department, you know we get them on the track towards becoming an electrical computer engineer. (UF Administrator)

This is an answer to the problem posed by administrators in chapter three. If students can join the engineering program in their first year, they can begin their early enculturation into engineering and begin the process of becoming an engineer.

Early Enculturation into Engineering

Students attracted to engineering programs through recruitment materials as well as from their own experiences with the application process and related activities may be frustrated by learning science and mathematics in environments outside the engineering context: "And that's what actually the students [say] after two years, 'What am I doing here? I'm studying math and science, that's not engineering.' That's one of the reasons. So they cannot relate" (FAMU-FSU administrator). Another administrator describes how engineers can build stronger relationships among their undergraduate student peers, something that is not possible during the first two years of the engineering program in most institutions because "they are detached from the real engineering world....And this is all on campus being taught by scientists on campus. They are not interacting with engineers." She goes on to explain:

> And there's a difference between science and engineering. For us as engineers, we always relate to things. So we teach formulas. We teach math. But we always relate to something, [not] stuff like the abstract way that they teach [elsewhere] on campus....That's why actually I would like to have these courses taught here in the college and being taught by Engineers because they can immediately...relate to the students, give them examples of real life and they actually get them out of that abstract kind of thing [that gets them] confused the first two years. (FAMU-FSU Administrator)

Early contact with undergraduate engineering programs fosters the development of an engineering identity for first and second year students who generally do not take their first engineering course until their junior year. On the other hand, as they are currently structured, engineering classes taken in the freshman and sophomore years are unlikely to engage students in active learning because most of these courses are taught as large lecture classes. Chapter four shows that professors we spoke with agreed that students are bored with the science and mathematics foundations of engineering, "They usually get weeded out in Calc III or Physics II or something like that" before they learn more about what engineers actually do. This administrator continues:

> A lot of students who are more team oriented think engineering will be too much isolation, and they don't really realize that actually [the] ability to function on teams is one of our major outcomes now...[and also to] ensure

that students can do team based engineering.... [It] is much more common than it ever has been, but we never get a chance to tell them that, so that's one of the things we're trying to put in [the Fundamentals of Engineering course], that if you like to work in teams, engineering is not necessarily a bad move. (UF administrator)

Administrators assert that students leave during their first two years because they do not understand what it means to be an engineer. An administrator contends, "The job basically at the pre-engineering level [is to get the students hooked in], not [waiting until] they become juniors and seniors... Once they become juniors and seniors you see them all the way until graduation (FAMU Administrator). One strategy for getting students into early and regular contact with engineering department faculty is to offer lower-level courses within the engineering department or college rather than elsewhere in the university. Engineering administrators extol the benefits of creating opportunities for undergraduates with an interest in engineering or who have declared a pre-engineering major to meet engineers in the field as early in their undergraduate coursework as possible. An obvious benefit of early enculturation is the development of an engineering identity from the outset.

Constructing New Pedagogical Frameworks

Pedagogy that employs real world applications as explained in chapter four and discussed in the next section of this chapter helps students learn practical engineering skills. This approach also develops their identities as engineers—it helps them to picture their futures as engineers in a concrete manner. When being an engineer remains abstract, or students do not self identify as future engineers, students may seek out other majors such as business they perceive as more tangible. When asked what could have been done to improve her experience in engineering, a switcher responded:

> I would definitely... have liked a class that they had to take that was more like interactive with what they're actually going to be doing so that they could figure it out their first year....This is how much you're going to get paid, these are the job opportunities out there... because I think that's really a good thing to know going into that program or any program at all. (USF Native American female switcher)

Several of the programs we visited do have introductory engineering courses, and we hold that these courses are important for students as they develop their identities. One student, when discussing such a program, thought that it could be improved if the course would present the engineering disciplines or subfields and also presented realistic guidance on how to succeed in them:

> I mean freshman lab is just discussing all the engineering disciplines. See what you want to pick from there, and I think they could do a better job of telling the demand that it's actually going to take on you as far as course work and studying time and that type of thing. (FAMU Black male student)

As such, these programs should not only expose students to the options for engineering careers, they should also outline the necessary steps to graduation.

Many professors we interviewed used words such as "interactive" and "engaging" to describe their pedagogical approaches. Our own observations in addition to student descriptions of their classes lead us to believe an interactive approach is actually not dominant in any of the programs we studied. Professors believe engineering courses are best suited to an approach that actually limits student interaction, namely a lecture-based format, supplemented by problem solving activities undertaken by professors themselves, at the board in front of the class.

Although they may make attempts to keep students engaged by asking questions, professors rarely use more active methods, such as providing opportunities for students to work problems while in class either individually or in small groups. Professors who did mention the importance of in-class group work said it was time-consuming and they could only pursue this strategy occasionally, suggesting they consider active methods supplemental rather than primary. Professors' reticence in using cooperative learning disproportionately impacts females and underrepresented minorities. Often women come to engineering lacking experience with mechanical activities. As such, many face challenges in relating their experiences to abstract theories (Felder et al., 1995). Both women and underrepresented minorities benefit from working in small groups; this approach helps in overcoming social isolation and reducing the fear of being singled out because of their small numbers (Henes et al., 1995; Murphy et al., 2007; Springer at al., 1999).

Real-life applications within the university. The need for more hands-on activities is a recurring theme among students when asked what could be improved about their programs: "If it wasn't for the lab a lot of kids would probably be lost because I would say I probably rely more on my lab than I do on class" (FAMU Black female student). We see a very practical need for labs to supplement classroom instruction. In addition, there is no question that working with professors on their research also serves the purpose of providing students access to the latest lines of inquiry in the field. Students recognize research as important to their programs of study: "Research, we have the most grant money I think in the nation for engineering research so I'd say that's definitely their strength" (USF White male student).

While this student's statement about USF's research dollars may be an exaggeration of the truth, what it reveals is the importance to students of the research enterprise. Students also see the advantage of undergraduate research collaboration with professors and suggest these collaborations become integrated into the curriculum: "I would recommend...more undergrad research...because I don't see much interaction with professors...unless it has to do specifically with the course they're taking." (FIU Hispanic male student). Another student comments:

> I think more labs are always good because that's where you really see, I mean you really don't get anything just learning concepts and reading the book, but when you see the applications is when you have a better understanding and when...you're learning what it's for, not just learning the concept." (FIU Hispanic male student)

In addition to conducting research with professors, students benefit from internships and placements in industry and similar settings outside of the university.

Real-life applications outside of the university setting. One student explained how his internship gave him a broader perspective than was the case for students who had to rely on coursework only to provide an understanding of the field:

> So I learned a whole lot...as far as contractors, building contractors, DOT work, how things get built...so this [capstone course] is...an orientation of what I learned this summer...... Which I guess is a good thing for somebody who hasn't had an internship. The pure engineering courses I'm fine with doing but I'm not enthusiastic about it because I am more interested in things that can be applied to, to a broader level, a better perspective. (UF White male student)

This student felt he was more prepared than other students in the class and was able to apply what he learned over the summer to his experience in the capstone engineering course. A student who left engineering also recommended connecting students to the engineering profession to encourage them to continue in their programs:

> If they could gather more of their students...and create or make some kind of council or some type of group that's really into focusing on engineering

projects within the school and outside in the Tampa Bay Community and maybe students could get involved with that and see engineering at work. Maybe seeing engineering at work would really inspire them to continue on and know that once I finish this sequence and once I finish this major, I can be off doing things like this all around the world. (USF Black female switcher)

With the goal of increasing persistence to engineering degrees, program administrators should listen to these student's suggestions to supplement the curriculum with outside activities. Other students also commented that internships provided them with mentors:

> I worked as an internship over the summer and a lot of them... really... they mentored me a lot over the summer because I didn't really know how, the secret of becoming... a PE and everything... and they told me about it and I still actually worked for them so if I ever have any questions I could email them and they would help me. (UF White female student)

Mentors are particularly important to female students, but can benefit all students when professionals in the field inspire them and share their strategies for success. We believe that each of these suggestions—more active learning in classrooms, more lab work including working with professors on their personal research programs, and internships with professional engineers—would strengthen the engineering pedagogical framework and encourage more students to complete degrees and pursue careers in engineering.

Personal Advising and Communication

Chapter six reveals how students feel some professors are not committed to teaching, and their departments care more about the prestige of their engineering program than individual student well-being and success. One student expressed her frustration with how FAMU-FSU was handling their Accreditation Board for Engineering and Technology (ABET) accreditation:

> It's just like they want to get the changes made. They don't really care about the students right now... I don't really like how our classes are being held right now and I'm glad that I'm graduating... They're trying to do a lot for the school and they're not caring about the students. (FAMU-FSU Black female student)

This student distinguishes between what faculty and administrators are doing for the school and actually caring about students. No doubt faculty and administrators had the best interests of students in mind as they worked for accreditation at FAMU-FSU. Faculty and administrators across universities also value quality teaching as discussed in chapter six. But students want more from faculty. They want support: faculty support was valued by the most students across universities suggesting that students want more than quality instruction. They want to be listened to and supported by faculty outside the classroom as well. They want to be understood and valued. In other words, students want to know faculty can help them become engineers both inside and outside the classroom. This contrast between student and faculty values may reflect findings that show engineering graduates appreciate the quality of their instruction less than students in other majors despite appreciating their major more than students in other majors (Bradburn, Nevill, & Cataldi 2006). The same student as above provides two examples of treatment she receives when voicing her complaints to faculty and administrators:

> And then... when we make a complaint about something, it's always that the students aren't capable of doing this... But when we try to become capable they don't want to help us... I know in one of my classes they're... doubling up on our work because they need to make requirements for ABET accreditation. So it's like they're doubling up on our work, but we don't have enough time to do the work. So... then we make a complaint about it, it's always that we're lazy or... we should be able to do this. (FAMU-FSU Black female student)

This student claims her complaints are rebuffed by faculty and administrators saying that students aren't capable or lazy or should know how to do the work. In some respects, this quote is an example of the strengths of this book. Taken by itself, researchers and readers could conclude that this student was merely exaggerating. Taken in context with interviews with faculty and administrators, it is clear faculty and administrators do feel that students are not doing everything they could be doing on their own to succeed. It is also possible that some faculty and administrators use the same harsh tone described above as they did in interviews.

Engineering students must be held to high academic standards; however, they do not necessarily understand what is expected by their professors and are likely to have difficulty seeking out faculty guidance, especially as freshmen and sophomores. When students identify faculty members whom they characterize as role models, it is generally because these individuals direct activities students participate in or because students see them as particularly, helpful, supportive, and accessible. Both faculty and students admit that it is up to students to seek out faculty, not the other way around.

> The University is large so... they can't make each person feel special... In the end I think it comes down to, are you willing to... put your own effort

into it. Because I don't think there's any one person that will sit down and carry you. (UF White male student)

Faculty believe students must take responsibility for their success, yet students remain frustrated by faculty members' perceived inability to communicate their expectations in a way that is understood by students. Advising provides an opportunity for students to articulate problems with course sequencing or course offerings. Many students believe both could be improved and advising sessions constitute a great opportunity for students to express frustration or make recommendations. We believe students should be advised by people trained for this purpose. A Black female switcher describes how advising is more than just relaying information; it is about students feeling supported and important:

> Once you make that initial contact with the student and you want them to be a part of your engineering program, really following up on what you say that you are going to do. Not just kind of sending them this information, we're interested in you, but once we kind of got you here, we don't really too much want that relation, because we got you to come here. (USF Black female switcher)

This switcher believes that once students are accepted, they can no longer rely on the university for support aside from an initial advising meeting. She recommends making this meeting not "something mandatory like we just come in [for] fifteen minutes and this is what you are going to take and everybody goes home." Instead, she believes the department should invest time and effort into "really finding out why that student wants to be an engineering student and making sure that they are on the right track because there are so many different...types of engineering."

Students should be counted on to take initiative in pursuing their own academic success, but they need continual guidance and direction throughout the time they spend in their programs. Even if personal advising and support is not available for all students every day, general communication should be provided routinely. One student wondered, "How do they get the word out? You know what I mean?....If they were to develop a great strategy and they didn't tell anybody, then it's not a great strategy" (UF White male student). Students often report a lack of transparency in the requirements for the programs they are pursuing. One student describes the difficulty he has preparing for the Fundaments of Engineering (*FE*) exam students must pass in order to become a licensed professional engineer.

> I know a girl...She got mad because...you have to register to take the test 6 months in advance in October...They post in front of the advising

office, but they don't really say it in class. They don't really tell you how the processing works. You have to kind of go on the website yourself and try. I mean, they talk about what's going to be on the test, but they don't tell you all the processes to apply for the test, so I think that's something the school could improve on. (FIU White male student)

Critical information like this may not reach every student because departments lack official channels of communication. Faculty and administrators may believe they are spreading the message, but students may not receive it. Especially critical is information on research opportunities for undergraduate students. As discussed throughout the book, students think their undergraduate experiences can be enhanced by improved access to research opportunities including opportunities to work in professors' labs and engage in internships. These positive collaborative experiences with faculty are key in student development, but these are opportunities that few students enjoy. Those students who do collaborate with faculty appreciate the straight-forward, somewhat stern approach used by some faculty. One student describes positive encouragement she receives from an administrator:

> He will not let you put excuses to anything you do. You have to take full responsibility of your actions... You have time here and you have time here and have time there... He's actually right so it makes you feel like you can do more.... He'll push you to be the best you can so that's one of the things that I really admire about him. (UF Hispanic female student).

She also describes how this administrator goes above and beyond his job description. He regularly invites members of student organizations to his house for dinner. He offered her a tutoring job when she was having financial troubles. He sent her resume out to potential employers. Stories like this from students are encouraging, but far too infrequent. Mentorship and support relies heavily on students choosing faculty members to mentor them and faculty in turn being willing to do so.

Race, Gender, and Mentorship

A key problem is that strong faculty-student relationships often depend on basic principles of friendship formation, among them "birds of a feather flock together." As chapter three shows, some Black switchers contend faculty are more likely to mentor students of the same race or nationality. Black students complained that they saw few if any professors who, "look like me." UF Black students in a focus group claimed that faculty diversity was "really lacking" with some saying "I've never had a female teacher" and "I've never had a minority teacher." Where do these students go for advising and mentorship?

A Black female in the focus group we just discussed reported said, "I definitely saw no women in my department [and] when I worked this summer? 40 men. And I was just an intern. And [they said] 'oh you're the first girl we've had in here in years'." Since women's early socialization may limit their exposure to STEM careers, it is particularly important for them to see other women engaged and successful in careers in these fields, as this provides implicit information about future professional possibilities (Zirkel, 2002). A UF female student described a positive relationship she has with a male administrator and her desire to be a female engineering role model in the future: "And I think I want to be like [the administrator] was when I'm already older and I want to be a mentor for other people who are younger that are in engineering" (UF Hispanic female student). Even though that male administrator has been a positive influence on her, she also described the impact of seeing a prominent female in engineering as a role model:

> I went to a conference this weekend and they were giving out awards because it was an award ceremony going on and they had this award for this lady...She was a[n] executive with IBM....Inside IBM she started a program where they actually have a support system for women that are engineers in IBM and that program has grown exponentially since she got there.... Right now she's the VP of finance in IBM. So she started that program and it has tons and tons of people in it and I don't know I just thought it was really cool because she started [as an] engineer, that's it. And then she moved up the chain and I thought it was really really cool and at the same time she was trying to be a mentor for people and I thought it was cool. (UF Hispanic female student)

This female engineering student admired a woman she saw receive an award at a conference, but when asked if there were any faculty members in her department she could identify with, she said there were none. While many women we spoke with describe the importance of having a mentor, none of them reported actually having a mentor. This may be due to the low percentages of women or minority faculty serving as mentors and also highlights the need for women and minorities to be introduced to engineers and engineering at very young ages. Such programs address the gap caused by less access to cultural capital before college that is typically the lot of women and underrepresented minority students in comparison with their white male peers.

A relationship involving a role model is defined as an unstructured relationship in which the individuals do not always have personal contact; however, while contact may be limited, the less-experienced individual admires and attempts to imitate the success of the more-experienced individual (Kram, 1985; Murray, 1991). We find that students wish to believe they can follow in the footsteps of their professors, and think that faculty who have experienced the process of becoming engineers should serve as guides and role models to their students. Reflecting this orientation, a switcher remarked:

> I would say listen to students. I mean, they want to learn, I wouldn't say everybody but the majority of them are here to learn and you know, don't just give them that they know, teach them, you know, teach them, reiterate it if you have to. I mean, I believe that's what professors are there for, they already went through everything, did their undergrad, grad, wherever they're at in their life now. I'm an undergrad and I want to get where you are, so help me... (USF Black female switcher)

This student recognizes the challenge of one-on-one time with faculty and suggests creating groups, or cohorts, within courses, "so students can follow, follow along." Programs through which women and minority students can see and identify with working engineers is important in affirming they can also become engineers.

Student Advising, Student Organizations and Fit

Students report having difficulty with advising; they are often unsure or dissatisfied about course sequencing; they believe advisors do not really care about them; and, finally, they question the information advisors provide. As a result, many students rely on other students for information, and while this may foster camaraderie and collaboration, students are left relying on hearsay or patently false information. A wise course of action for undergraduate programs to pursue is one that gives students access to the most reliable information, not easily accomplished because the best source for this information may be faculty members who are already stretched with multiple demands on their time.

The fact remains, however, that most students report struggling with the advising they receive. They are often unsure or dissatisfied about course sequencing. They believe advisors do not really care about them and question the information advisors provide. Instead, student organizations fill a void for students and should be integrated into institutional support system. Faculty, especially faculty advisors, often use student organizations to relay crucial information to students they may not receive through other official channels.

Similarly, students benefit from peer mentoring programs. Engineering students rely heavily on the advice they receive from students, a form of social capital, particularly when they believe they are not getting support or guidance from their programs. One former military transfer student from a community college to a four-year program likened the institutional knowledge of older students to military intelligence—students who have been in the program "know the ropes," that is, they have the cultural capital supporting what it takes to succeed in the program. Peer mentoring programs also link students to student organizations useful for all students but particularly for women and minority students.

Each chapter in this volume provides examples underscoring the importance of institutional and program "fit." Why is "fit" so important? If students do not sense their program of study and their selected engineering pathway suits them, they will not remain in engineering. There is some advantage to being a male, a "nerd," or an international student because one's status or fit with other students builds camaraderie and a sense of belonging. Developing an engineering identity and self identifying as an engineer is critical to succeeding in such a rigorous and challenging program. During the course of our interviews, Black students and women mention their heavy reliance on minority and women-centered student organizations, for example, the National Society of Black Engineers (NSBE) and the Society for Women Engineers (SWE). These organizations serve as academic and social support systems: students feel connected to students they identify with, and also learn important skills to succeed in engineering- skills they might not have entered the programs with- such as understanding the importance of research or knowing how to approach faculty. The acquisition of these skills makes it easier for students to earn the social capital necessary to fit in with their engineering classmates and develop an identity as an engineering, an identity that aids in persistence toward an engineering degree.

The Importance of Capital to Persistence

While we do not disregard the negative effects of differential math preparation, early interest in engineering, pedagogy, or any other factor critical to success in engineering as discussed in this volume, we also argue that the difference between switchers and persisters is not simply discrepancies in high school preparation. Regardless of preparation, engineering curricula challenge persisters who develop strategies to succeed. Chapter three shows much of the frustration administrators and staff experience in counseling students to develop the mentality to persist. Kunda (2006) remarks that engineering culture resonates with a "general demeanor" combining a "studied informality, a seemingly self-assured sense of importance, and a clearly conveyed impression of hard, involving, and strangely enjoyable, even addictive work"(p. 2).

Administrators and faculty expect successful students to naturally develop this same demeanor and gradually to become addicted to engineering work. An administrator described how switchers lack the desire to persist necessary in order to fulfill all dimensions of the engineering role:

> [Switchers] go, "Oh my gosh, I guess I can't do engineering" and then they go find something they're immediately successful in. And engineering is not rocket science. Engineering is the result of persistence and persistence and persistence. And it requires a person, that if they flunk calculus, they take it again, and then they take it again. And those make really good engineers, but they don't see that. (USF administrator)

Switchers may leave engineering not because they are incapable of completing the requirements, but because they have not learned from faculty, advisors, upper-class students, peers, family and friends, or their own experiences what is needed to persist to earn an engineering degree. Whether it is proper study habits or reacting to a poor grade in a course or how to ask for help or how to retake engineering courses when they fail, students need to know what to expect from the engineering courses and experiences they will have as undergraduates and specifically what they need to do to ensure their graduation. Faculty, administrators, and staff talk about how students do not realize this; however, it is clear that these individuals must embrace their roles as mentors and learn how to effectively communicate with students in a way students can understand.

Students who persist see these expectations, embrace the struggle, and take time to develop into their own version of becoming an engineer: "I guess if it was easy then we'd all be engineers.... We want it to be a good program where it's actually going to make you learn something" (USF White male student). Persistence is the result of a genuine, enduring interest in engineering, the resources to develop strategies to persist, and the will to persist and overcome failure to do so. These are key elements to success above and beyond mathematics and science preparation or general aptitude and ability. All students regardless of the skills and training they bring into engineering must work hard to learn the engineering culture of persistence from institutional gatekeepers—faculty, administrators, and staff—who are willing to patiently teach and mentor them. It is this persistence capital that has the most salience within the culture and climate of engineering programs.

Note

1. It is inappropriate to generalize from our findings to make recommendations for programs in elite institutions, although some findings seem to cut across many if not all types of engineering programs.

References

Adelman, C. (2004). *Principal Indicators of Student Academic Histories in Postsecondary Education, 1972–2000.* Washington, DC: ED Pubs.

——— (2006). *The Toolbox Revisited: Paths to Degree Completion from High School through College.* Washington, DC: U.S. Dept. of Education.

Allan, E. J., & Madden, M. (2006). Chilly Classrooms for Female Undergraduate Students: A Question of Method? *Journal of Higher Education, 77*(4), 684–711.

Aparicio, A. (2007). Contesting Race and Power: Second-Generation Dominican Youth in the New Gotham. *City & Society, 19*(2), 179–201.

Appadurai, A. (2001). Grassroots Globalization and the Research Imagination. In A. Appadurai (Ed.), *Globalization.* Durham, NC: Duke University Press.

Astin, A.W. (1984). Student Involvement: A Developmental Theory for Higher Education. *Journal of College Student Personnel, 25*(4), 297–308.

——— (1985). *Achieving Educational Excellence.* San Francisco, CA: Jossey-Bass.

——— (1993). *What Matters in College?: Four Critical Years Revisited.* San Francisco, CA: Jossey-Bass.

——— (1999). Student Involvement: A Developmental Theory for Higher Education. *Journal of College Student Development, 40*(5), 518.

Attinasi, L.C., Jr. (1989). Getting In: Mexican Americans' Perceptions of University Attendance and the Implications for Freshman Year Persistence. *Journal of Higher Education, 60*(3), 247–277.

Berger, J.B. (2002). Understanding the Organizational Nature of Student Persistence: Empirically-based Recommendations for Practice. *Journal of College Student Retention: Research, Theory & Practice, 3*(1), 3–21.

BEST (2004). *A Bridge for All: Higher Education for All: Higher Education Design Principles to Broader Participation in Science, Technology, Engineering, and Math, Building Engineering and Science Talent.* San Diego, CA: Best.

Bleeker, M.M., & Jacobs, J.E. (2004). Achievement in Math and Science: Do Mothers' Beliefs Matter 12 Years Later? *Journal of Educational Psychology, 96*(1), 97–109.

Bonous-Hammarth, M. (2000). Pathways to Success: Affirming Opportunities for Science, Mathematics, and Engineering Majors. *Journal of Negro Education, 69*(1–2), 92–111.

Borman, K.M., Cotner, B.A., Lee, R.S., Boydston, T.L., & Lanehart, R.E. (2009). *Improving Elementary Science Instruction and Student Achievement: The Impact of a Professional Development Program.* Paper presented at the American Educational Research Association.

Borman, W. (1990). Job Behavior, Performance and Effectiveness. In M. Dunnette & L. Hough (Eds.), *Handbook of Industrial-Organizational Psychology*. Palo Alto, CA: Consulting Psychologists Press.

Bourdieu, P. (1977). *Outline of a Theory of Practice*. New York: Cambridge University Press.

Brady, K.L., & Eisler, R.M. (1999). Sex and Gender Equity in the College Classroom: A quantitative Analysis of Faculty-Student Interactions and Perceptions. *Journal of Educational Psychology, 9*, 127–145.

Brainard, S.G., & Carlin, L. (1998). A Six-Year Longitudinal Study of Undergraduate Women in Engineering and Science. *Social Education, 62*(7), 369–376.

Braxton, J.M., & McClendon, S.A. (2002). The Fostering of Social Integration and Retention through Institutional Practice. *Journal of College Student Retention: Research, Theory & Practice, 3*(1), 57–71.

Bryant, A.N. (2001). ERIC Review: Community College Students: Recent Findings and Trends. *Community College Review, 29*, 77–93.

Buchmann, C., & DiPrete, T.A. (2006). The Growing Female Advantage in College Completion: The Role of Family Background and Academic Achievement. *American Sociological Review, 71*(4), 515.

Calcagno, J.C., Crosta, P., Bailey, T., & Jenkins, D. (2007). Stepping Stones to a Degree: The Impact of Enrollment Pathways and Milestones on Community College Student Outcomes. *Research in Higher Education, 48*(7), 775–801.

Campbell, P.B., Jolly, E., Hoey, L., & Perlman, L.K. (2002). *Upping the Numbers: Using Research-based Decision Making to Increase Diversity in the Quantitative Disciplines*. Groton, MA: Campbell-Kibler Associates.

Caporrimo, R. (2008). Community College Students: Perceptions and Paradoxes. *Community College Journal of Research and Practice, 32*(1), 25–37.

Caso, R., Clark, C., Froyd, J., Inam, A., Kenimer, A., Morgan, J. et al. (2002). A Systemic Change Model in Engineering Education and Its Relevance for Women. *American Society for Engineering Education Conference Proceedings*.

CEOSE (2002). *Committee on Equal Opportunities in Science and Engineering*. Washington, DC.

Chen, H.L., Lattuca, L.R., & Hamilton, E.R. (2008). Conceptualizing Engagement: Contributions of Faculty to Student Engagement in Engineering. *Journal of Engineering Education, 97(3)*, 339–353.

Clewell, B.C., & Campbell, P.B. (2002). Taking Stock: Where We've Been, Where We Are, Where We're Going. *Journal of Women and Minorities in Science and Engineering, 8*, 255–284.

Darling-Hammond, L. (1995). Inequality and Access to Knowledge. In C.A. B.J.A. Banks (Ed.), *Handbook of Research on Multicultural Education*. New York: Macmillan.

Dirks, N.B., Eley, G., & Ortner, S.B. (1998). *Culture/Power/History: A Reader in Contemporary Social Theory*. Princeton, NJ: Princeton University Press.

Faulkner, W. (2000). The Power and the Pleasure? A Research Agenda for "Making Gender Stick" to Engineers. *Science, Technology, & Human Values, 25(1)*, 87–119.

Felder, R.M. (1995). A Longitudinal Study of Engineering Student Performance and Retention. Part IV. Instructional Methods and Student Responses to Them. *Journal of Engineering Education, 84(4)*, 361–367.

Foucault, M. (1980a). *The History of Sexuality, Vol. I. An Introduction* New York: Vintage.
——— (Ed.). (1980b). *Power/Knowledge: Selected Interviews and Other Writings, 1972–1977* New York: Pantheon.
Frestedt, J.L. (1995). Mentoring Women Graduate Students: Experience of the Coalition of Women Graduate Students at the University of Minnesota, 1993–1995. *Journal of Women and Minorities in Science and Engineering, 2*, 151–170.
Friedman, T.L. (2005). *The World Is Flat: A Brief History of the Twenty-first Century.* New York: Farrar, Straus and Giroux.
García Canclini, N. (1989). *Hybrid Cultures: Strategies for Entering and Leaving Modernity.* Minneapolis, MN: University of Minnesota Press.
Giddens, A. (1979). *Central Problems in Social Theory: Action, Structure, and Contradiction in Social Analysis.* Berkeley, CA: University of California Press.
Glick Schiller, N., Basch, L., & Szanton Blanc, C. (1995). From Immigrant to Transmigrant: Theorizing Transnational Migration. *Anthropological Quarterly, 68*(1), 48–63.
Glick, W.H. (1985). Conceptualizing and Measuring Organizational and Psychological Climate: Pitfalls in Multilevel Research. *The Academy of Management, 10*(3), 601–616.
Gokhale, A.A., & Stier, K. (2004). Closing the Gender Gap in Technical Disciplines: An Investigative Study. *Journal of Women and Minorities in Science and Engineering, 10*, 149–159.
Goodman, I.F., Cunningham, C.M., Lachapelle, C., Thompson, M., Bittinger, K., Brennan, R.T. et al. (2002). *Final Report of the Women's Experiences in College Engineering.* Cambridge, MA: Goodman Research Group.
Gramsci, A. (1971). Selections from the Prison Notebooks of Antonio Gramsci. Ed. Nowell-Smith & Q. Hoare. New York: International Publishers.
Hall, R.M., & Sandler, B.R. (1982). *Out of the Classroom: A Chilly Campus Climate for Women?* Washington, DC: Association of American Colleges.
Halperin, R.H. (1994). *Cultural Economies Past and Present.* Austin, TX: University of Texas Press.
Harvey, D. (1989). *The Condition of Postmodernity: An Inquiry into the Origins of Culture Change.* Cambridge, MA: Blackwell.
Henes, R., Bland, M.M., Darby, J., & McDonald, K. (1995). Improving the Academic Environment for Women Engineering Students through Faculty Workshops. *Journal of Engineering Education, 84(1)*, 59–67.
Horn, L., National Center for Education, S., & Institute of Education, S. (2006). *Placing College Graduation Rates in Context: How 4-Year College Graduation Rates Vary with Selectivity and the Size of Low-Income Enrollment: Postsecondary Education Descriptive Analysis Report.* Washington, DC: U.S. Department of Education.
Howe, N., & Strauss, W. (2000). *Millennials Rising: The Next Great Generation.* New York: Vintage Books.
Huang, G., Taddese, N., & Walter, E. (2000). *Entry and Persistence of Women and Minorities in College Science and Engineering Education.*
Irvine, J.J., & York, D.E. (1995). Learning Styles and Culturally Diverse Students: A Literature Review. In J.A. Banks (Ed.), *Handbook of Research on Multicultural Education.* New York: Simon & Schuster Macmillan.

James, L.R. (1982). Aggregation bias in estimates of perceptual agreement. *Journal of Applied Psychology, 67*, 219–229.

James, L.R., Demaree, R.G., & Wolf, G. (1984). Estimating Within-Group Interrater Reliability with and without Response Bias. *Journal of Applied Psychology, 69*, 85–98.

――― (1993). r_{wg}: An assessment of within-group interrater agreement. *Journal of Applied Psychology, 78*, 306–309.

Jones, L., Castellanos, J., & Cole, D. (2002). Examining the Ethnic Minority Student Experience at Predominantly White Institutions: A Case Study. *Journal of Hispanic Higher Education, 1*, 19–39.

Klein, K.J., & Kozlowski, S.W.J. (2000). A Multilevel Approach to Theory and Research in Organizations: Contextual, Temporal, and Emergent Processes. In K.J. Klein & S.W.J. Kozlowski (Eds.), *Multilevel Theory, Research and Methods in Organizations* (pp. 3–90). San Francisco, CA: Jossey-Bass.

Kraemer, B.A. (1997). The Academic and Social Integration of Hispanic Students into College. *Review of Higher Education, 20*(2), 163–180.

Kristof, A.L. (1996). Person-Organization Fit: An Integrative Review of its Conceptualizations, Measurement, and Implications. *Personnel Psychology, 49*(1), 1.

LeCompte, M.D., & Schensul, J.J. (1999). *Designing & Conducting Ethnographic Research*. Walnut Creek, CA: AltaMira Press.

Lee, V.E., Mackie-Lewis, C., & Marks, H.M. (1993). Persistence to the Baccalaureate Degree for Students Who Transfer from Community College. *American Journal of Education, 102*(1), 80.

Leech, N.L., & Onwuegbuzie, A.J. (2009a). A Proposed Fourth Measure of Significance: The Role of Economic Significance in Educational Research. *Evaluation & Research in Education, 18*(3), 179.

――― (2009b). A Typology of Mixed Methods Research Designs. *Qual. Quant. Quality and Quantity, 43*(2), 265–275.

Lewin, K. (1951). *Field Theory in Social Science; Selected Theoretical Papers*. New York: Harper.

Lindell, M.K., & Brandt, C.J. (2000). Climate Quality and Climate Consensus as Mediators of the Relationship between Organizational Antecedents and Outcomes. *Journal of Applied Psychology, 85*(3), 331–348.

Locks, A.M., Hurtado, S., Bowman, N.A., & Oseguera, L. (2008). Extending Notions of Campus Climate and Diversity to Students' Transition to College. *Review of Higher Education, 31*(3), 257–286.

Margolis, J.a. F.A. (2002). *Unlocking the Clubhouse: Women in Computing*. Cambridge, MA: MIT Press.

Mattis, M.C., & Sislin, J. (2005). *Enhancing the Community College Pathway to Engineering Careers*. Washington, DC: National Academies Press.

May, G.S., & Chubin, D.E. (2003). A Retrospective on Undergraduate Engineering Success for Underrepresented Minority Students. *Journal of Engineering Education, 92*, 27–40.

Mellström, U. (1995). *Engineering Lives: Technology, Time and Space in a Male-Centred World*. Linkoping: Linkoping University.

Micceri, T. (2001). *Change Your Major and Double Your Graduation Chance*. Paper presented at the Association for Institutional Research Annual Forum.

Micceri, T., & Wajeeh, E. (1998). *The Influence of Geographical Location and Application to Multiple Institutions on Recruitment.* Paper presented at the Association for Institutional Research Annual Forum.

Moore III, J.L. (2006). A Qualitative Investigation of African American Males' Career Trajectory in Engineering: Implications for Teachers, School Counselors, and Parents. *Teachers College Record, 108*(2), 246–266.

Murphy, M.C., Steele, C.M., & Gross, J.J. (2007). Signaling Threat: How Situational Cues Affect Women in Math, Science, and Engineering Settings. *Psychological Science, 18*(10), 879–885.

Naumann, S.E., & Bennett, N. (2000). A Case for Procedural Justice Climate: Development and Test of a Multilevel Model. *Academy of Management Journal, 43*, 881–889.

Neal, A., & Griffin, M.A. (2006). A Study of the Lagged Relationships among Safety Climate, Safety Motivation, Safety Behavior, and Accidents at the Individual and Group Levels. *Journal of Applied Psychology, 91*, 946–953.

Nixon, A.E., Meikle, H., & Borman, K. (2007). The Urgent Need to Encourage Aspiring Engineers: Effects of College Degree Program Culture on Female and Minority Student STEM Participation. *Latin American and Caribbean Journal of Engineering Education, 1*(2), 57–63.

Noel, L., Levitz, R., & Saluri, D. (1985). *Increasing Student Retention: Effective Programs and Practices for Reducing the Dropout Rate.* San Francisco, CA: Jossey-Bass.

NSB (2002). *Science and Engineering Indicators—2002.* Arlington, VA: National Science Foundation.

——— (2008). *Science and Engineering Indicators 2008.* Arlington, VA: National Science Foundation.

NSF (1999). *Women, Minorities, and Persons with Disabilities in Science and Engineering: 1998.* Arlington, VA: National Science Foundation.

Nye, B., Hedges, L.V., & Konstantopoulos, S. (2000). The Effects of Small Classes on Academic Achievement: The Results of the Tennessee Class Size Experiment. *American Educational Research Journal, 37*(1), 123.

OPPAGA, (2006). *Most Acceleration Students Perform Well, but Outcomes Vary by Program Type.* Tallahassee, FL: Office of Florida Legislature.

O'Reilly, C.A., Chatman, J., & Caldwell, D.F. (1991). People and Organizational Culture: A Profile Comparison Approach to Assessing Person-Organization Fit. *Academy of Management Journal, 34*(3), 487–516.

Ostroff, C.L., & Schulte, M. (2007). Multiple Perspectives of Fit in Organizations across Levels of Analysis. In C.L. Ostroff & T. Judge (Eds.), *Perspectives on Organizational Fit.* New York: Lawrence Erlbaum Associates.

Ostroff, C., Kinicki, A.J., & Tamkins, M.M. (2003). Organizational Culture and Climate In W.C. Borman, D.R. Ilgen & R.J. Klimoski (Eds.), *Comprehensive Handbook of Psychology, Volume 12: I/O Psychology.* New York: John Wiley & Sons.

Pascarella, E.T., & Terenzini, P.T. (2005). *How College Affects Students: A Third Decade of Research.* San Francisco, CA: Jossey-Bass.

Penick, B.E., Morning, C., & National Action Council for Minorities in Engineering (1983). *The Retention of Minority Engineering Students: Report on the 1981–82 NACME Retention Research Program.* New York: National Action Council for Minorities in Engineering.

Perna, L.W., & Titus, M.A. (2005). The Relationship between Parental Involvement as Social Capital and College Enrollment: An Examination of Racial/Ethnic Group Differences. *Journal of Higher Education, 76*(5), 485–518.
Postill, J. (Forthcoming). Introduction: Theorising Media and Practice. In B. Bräuchler & J. Postill (Eds.), *Theorising Media and Practice*. New York: Berghahn.
Rendon, L.I., Jalomo, R.E., & Nora, A. (2000). Theoretical Considerations in the Study of Minority Student Program Efficacy in Higher Education. In J.M. Braxton (Ed.), *Reworking the Student Departure Puzzle* (pp. 127–156). Nashville, TN: Vanderbilt University Press.
Robbins, D. (2005). The Origins, Early Development and Status of Bourdieu's Concept of "Cultural Capital." *British Journal of Sociology, 56*(1), 13–30.
Salter, D. W. (2003). Exploring the "Chilly Classroom" Phenomenon as Interactions between Psychological and Environmental Types. *Journal of College Student Development, 44*(1), 110–121.
Salvaggio, A.N., Schneider, B., Nishii, L.H., Mayer, D.M., Ramesh, A., & Lyon, J.S. (2007). Manager Personality, Manager Service Quality Orientation, and Service Climate: Test of a Model. *The Journal of Applied Psychology, 92*(6), 1741–1750.
Sandler, B.R., Silverberg, L.A., Hall, R.M., & National Association for Women in Education (1996). *The Chilly Classroom Climate: A Guide to Improve the Education of Women*. Washington, DC: National Association for Women in Education.
Sassen, S. (1998). *Globalization and Its Discontents*. New York: New Press.
Scheld, S. (2007). Youth Cosmopolitanism: Clothing, the City and Globalization in Dakar, Senegal. *City & Society, 19*(2), 232–253.
Schneider, B. (1990). The Climate for Service: An Application of the Climate Construct. In B. Schneider (Ed.), *Organizational Climate and Culture* (pp. 383–412). San Francisco, CA: Jossey-Bass.
Schneider, B., & Bowen, D.E. (1985). Employee and Customer Perceptions of Service in Banks: Replication and Extension. *Journal of Applied Psychology, 70*, 423–433.
Schneider, B., White, S.S., & Paul, M.C. (1998). Linking Service Climate and Customer Perceptions of Service Quality: Test of a Causal Model. *The Journal of Applied Psychology, 83*(2), 150–163.
Seymour, E., & Hewitt, N.M. (1997). *Talking about Leaving: Why Undergraduates Leave the Sciences*. Boulder, CO: Westview Press.
Smith, D.G., Gerbick, G.L., & Figueroa, M.A. (1997). *Diversity Works: The Emerging Picture of How Students Benefit*. Washington, DC: Association of American Colleges and Universities.
Springer, L., Stanne, M.E., & Donovan, S.S. (1999). Effects of Small-Group Learning on Undergraduates in Science, Mathematics, Engineering, and Technology: A Meta-Analysis. *Review of Educational Research, 69*(21), 21–51.
Steele, C.M., & Aronson, J. (1995). Stereotype Threat and the Intellectual Test Performance of African Americans. *Journal of Personality and Social Psychology, 69*(5), 797.

Tate, W.F. (2008). The Political Economy of Teacher Quality in School Mathematics: African American Males, Opportunity Structures, Politics, and Method. *American Behavioral Scientist, 51*(7), 953–971.

Tierney, W. (1992). An Anthropological Analysis of Student Participation in College. *Journal of Higher Education, 63*(6), 603–618.

Tinto, V. (1975). Dropout from Higher Education: A Theoretical Synthesis of Recent Research. *Review of Educational Research, 45*(1), 89–125.

——— (1993). *Leaving College: Rethinking the Causes and Cures of Student Attrition.* Chicago, IL: University of Chicago Press.

Tonso, K.L. (2006). Student Engineers and Engineer Identity: Campus Engineer Identities as Figured World. *Cultural Studies of Science Education, 1,* 273–307.

Tsapogas, J. (2004). *The Role of Community Colleges in the Education of Recent Science and Engineering Graduates.* Washington, DC: National Science Foundation.

Tyson, W. (2002). Understanding the Margins: Marginality and Social Segregation in Predominantly White Universities. In I. Robert M. Moore (Ed.), *The Quality and Quantity of Contact: African Americans and Whites on College Campuses* (pp. 307–322). Lanham, MD: University Press of America.

Tyson, W., Lee, R., Borman, K.M., & Hanson, M.A. (2007). Science, Technology, Engineering, and Mathematics (STEM) Pathways: High School Science and Math Coursework and Postsecondary Degree Attainment. *Journal of Education for Students Placed at Risk, 12*(3), 243–270.

Venezia, A., Venezia, A., Callan, P., Finney, J., Kirst, M., & Usdan, M. (2005). *The Governance Divide: A Report on a Four-State Study on Improving College Readiness and Success.* Washington, DC: National Center for Public Policy and Higher Education.

Vogt, C.M. (2008). Faculty as a Critical Juncture in Student Retention and Performance in Engineering Programs. *Journal of Engineering Education, 97(1),* 27–36.

Wallerstein, I. (1974). *The modern World System: Capitalist Agriculture and the Origins of the European World-Economy in the Sixteenth Century.* New York: Academic Press.

Willis, P. (1977). *Learning to Labour: How Working Class Kids Get Working Class Jobs.* Westmead, UK: Saxon.

Wyer, M. (2003). Intending to Stay: Images of Scientists, Attitudes toward Women, and Gender as Influences on Persistence Among Science and Engineering Majors. *Journal of Women and Minorities in Science and Engineering, 9,* 1–16.

Yang, J., Mossholder, K.W., & Peng, T.K. (2007). Procedural Justice Climate and Group Power Distance: An Examination of Cross-Level Interaction Effects. *The Journal of Applied Psychology, 92*(3), 681–692.

York, C.M., & Tross, S.A. (1994). *Evaluation of Student Retention Programs: An Essential Component.* Paper presented at the First Annual SUCCEED Conference on Improvement of Engineering Education.

Contributors

Kathryn M. Borman is Professor of Anthropology and lead researcher at the Alliance for Applied Research in Anthropology and Education, Department of Anthropology, University of South Florida. Her program of research has focused on urban school districts with high levels of minority students, implementing a science professional development program for elementary teachers to increase science achievement, and understanding STEM pathways to postsecondary education. Upcoming research includes projects funded by the National Science Foundation, the Department of Education, and the Spencer Foundation.

Susan Chanderbhan Forde is a doctoral candidate in the School Psychology program at the University of South Florida. She earned her B.A. in Psychology at York University in Toronto, Canada and her M.S. in Educational Psychology at Indiana University-Bloomington. She has worked in the nonprofit sector and in higher education administration in New York City and public schools in Indiana and Florida. Her primary research interests are the relationship between educational institutions and immigrant families and students and educational initiatives in developing countries. Currently, she is an intern in the Dallas Independent School District's Psychological Services Department.

Bridget A. Cotner is an educational anthropologist and qualitative researcher pursuing a doctoral degree in the Measurement, Research, and Evaluation Program in the College of Education, University of South Florida. She received her B.A. in Anthropology from Ball State University, and her M.A. in Applied Anthropology from the University of South Florida in 2001. Her research interests include working with practitioners to understand how school reform and accountability influence practice in the classroom and school.

Cynthia A. Grace is an environmental anthropologist and doctoral candidate in Applied Anthropology at the University of South Florida. She received her B.A. in Psychology from Wake Forest University, and her M.A. in Marine Affairs and Policy from the University of Miami's

Rosenstiel School of Marine and Atmospheric Science. After working as a graduate research assistant for two years with the Alliance for Applied Research in Education and Anthropology (AAREA), Ms. Grace moved to St. Croix, U.S. Virgin Islands to complete her dissertation research regarding fisheries management. Her dissertation examines the challenges facing fisheries management in St. Croix, including how to balance the need to conserve marine resources with the need to allow humans to continue using those resources.

Rhoda H. Halperin, Ph.D., received her B.A. in Economics from Bennington College and a Ph.D. in Anthropology from Brandeis University. Most recently, Dr. Halperin was Professor of Anthropology at Montclair State University and was Professor Emerita in the Department of Anthropology, University of Cincinnati. In 2006 Dr. Halperin became a Fellow of the Society for Applied Anthropology, recognizing her work on urban development, education and the theory and practice of community heritage. Her other research interests includes relationships between theory and practice in urban anthropology, gender and power, medical anthropology and cultural competency, and history and contemporary anthropological theory.

Rebekah S. Heppner is an applied cultural anthropologist and postdoctoral research scholar in the department of anthropology at the University of South Florida. Dr. Heppner holds an MBA from the University of Tampa and a B.S. from Oklahoma State University. She received her Ph.D. in Applied Anthropology from the University of South Florida in 2007. Her dissertation research comprised life history interviews of women executives in corporate environments. She is currently involved in qualitative research in the secondary education system in Florida.

Reginald S. Lee is a Ph.D. Candidate in Curriculum and Instruction with dual concentrations in Educational Measurement and Research, and Special Education at the University of South Florida. He received a M.A. in Special Education from the University of South Florida and a B.A. from McDaniel College in Sociology. He has examined the impact of systemic reform on mathematics and science achievement in urban schools and studied student engagement during instructional activities. His research interests include equity and access issues in public education with a current emphasis on high school and college course-taking and student achievement and attainment in science, technology, engineering, and mathematics.

Jason E. Miller is a doctoral candidate in Applied Anthropology at the University of South Florida where he is also concurrently pursuing a

Masters of Public Health in Health Education. Miller's research focuses on participatory and visual approaches to education, health, and community development. Miller is currently an instructor in the Department of Anthropology. Previously, Miller taught Anthropology at a community college in Oregon where he also coordinated the Multicultural Center.

Arland Nguema Ndong is currently enrolled in the doctoral program of the Department of Anthropology at the University of South Florida. Arland graduated from Université Omar Bongo (Gabon) with a B.A. in English and from University of Paris (France) with a M.A. in U.S. History and Civilization. His research interests include understanding how Internet use as a marketing tool in attracting students can contribute to increasing enrollments in science and engineering (S&E) departments, especially among women and minority students.

Chrystal A.S. Smith is a Ph.D. Candidate in Applied Biological Anthropology in the Department of Anthropology, University of South Florida. She received her B.A. in Anthropology from Howard University, M.A. in Applied Anthropology from the University of Maryland, College Park and M.P.H. with an emphasis on Epidemiology from the University of South Florida. Ms. Smith is the Project leader of the evaluation study of the St. Louis Center for Inquiry in Science Teaching and Learning (CISTL). Her research interests include population genetics, the development and implementation of health prevention and treatment policy targeting marginalized populations, the impact of chronic and infectious diseases in the Caribbean as well as the impact of diabetes and hypertension on the health of immigrant, minority, and low income communities in the United States.

Will Tyson, Ph.D. is an Assistant Professor of Sociology and senior research associate at the Alliance for Applied Research in Anthropology and Education, Department of Anthropology, University of South Florida. He has a Ph.D. in Sociology and a graduate certificate in Women's Studies from Duke University and a B.A. in Sociology and Psychology from Wake Forest University. He came to the University of South Florida in 2004 as an NSF Postdoctoral Research Associate and joined USF Sociology in 2005. His research interests include high school science and mathematics course-taking and factors that lead students into STEM degree attainment and careers and structural factors that enhance interracial contact and lead to interracial friendships.

Hesborn O. Wao, Ph.D. is a senior research associate at the Alliance for Applied Research in Education and Anthropology (AAREA) and a graduate in Measurement and Evaluation in 2008 from the University of South Florida (USF). He has a B.Ed. (Mathematics) degree from the University of Nairobi in Kenya and a M.Ed. (Measurement & Evaluation) from USF. Dr. Wao's research focuses on the effects of social organizations on student outcomes. His interests include application of mixed methods approaches to research/evaluation design; measurement of student outcomes; and secondary data analysis using advanced modeling techniques. He currently serves as a Research Associate, managing a five-year federally funded project, *"Evaluation of Florida Voluntary Public School Choice (VPSC) Programs."*

Cassandra Workman Whaler is a Ph.D. candidate in Applied Medical Anthropology in the Department of Anthropology and a M.P.H. student in Epidemiology at the University of South Florida. She received a B.A. in Anthropology from Northern Illinois University as well as a M.A. in Anthropology from Western Michigan University. She has spent the past four years working at the Alliance for Applied Research in Education and Anthropology examining gender and racial/ethnic parity in science, technology, engineering, and mathematics in higher education. Her additional research interests include HIV/AIDS, water insecurity, and women's health, specifically in Southern Africa. She completed an internship for the U.S. Agency for International Development in Mozambique in fall of 2009. She is currently set to begin dissertation research in Lesotho in spring of 2010.

Index

*bold denotes tables or figures

academic capital *see* capital
academic disciplines
  anthropology, 4,5,12, 15, 26–27, 105, 127, 152
  economics, 4
  industrial-organizational (I/O) psychology, 6–7, 12, 15–16, 57, 105, 130
  measurement, 12
  sociology, 5, 12
advanced placement, 87–88, 91, 103
advising, 118, 119, 143, 145, 168, 169, 170, 185, 188
Alliance for Applied Research in Anthropology and Education (AAREA), 9
Astin, Alexander, 8, 67, 106, 123
  involvement process, 107

Bourdieu, Pierre (1930–2002), 3, 5, 56, 102, 107, 124
  capital *see* capital
  habitus, 107, 124
  practice theory *see* theoretical frameworks

campus ecology, 7, 17, 21, 26, 48, 50–51
campus resources *see* campus ecology
capital, 6, 79, 128, 189
  academic, 54, 143, 178
  cultural, 5, 54, 55, 56, 64, 73, 75, 79, 82, 83, 85, 100, 102, 104, 105, 125, 128, 144–146, 153, 162, 163, 170–171, 175, 187, 189
  financial, 6, 54
  persistence, 190
  social, 5, 6, 55, 79, 104, 124, 125, 143, 153, 162, 170–171, 189
  symbolic, 5, 54, 55, 56, 145
chilly climate, 8, 107, 124–126, 153
Chubin, Daryl, 129
climate
  advising and mentoring, 118, 119, 120
  classroom interaction, 118
  climate for, 106
  collaboration, 106, 107
  communication, 120
  conceptual framework, 3–4, **3**
  course sequencing, 120
  definition of, 7–8, 15–16, 105
  engineering department climate, 4, 11, 40, 43–44, 47–48, 75–76, 125
  engineering organizations, 106, 107, 125 *see also* student organizations
  faculty support, 110, 112–118, **113**, **115**, **117**, 121, 125
  fit, 105–106, 111, 112, 117, 118, 122–125
  group work, 122
  I/O psychology, 12, 16, 49–50, 105 *see also* academic disciplines

climate—*Continued*
  institutional support, 110, 112, 116, 118–119
  methods for assessing, 17, 105, 107–109
  office hours, 118–119, *see also* office hours
  personal agency and peer support, 105, 116–118, **117**, 120–123 *see also* collaboration
  practice theory, 107
  qualitative component, 105, 108–109
  quantitative analysis, 111–112
  role models, 110, **117**, 119
  scales, 109–111, *see under individual scale names*
  socio-emotional support, 121–123
  staff support, 119
  survey, 12, 13, 16, 18, 108
  survey analysis, 15, 111–112
  survey measures, 109–11, *see under individual scale names*
  survey recruitment, 13
  survey results, 84, 86, 102, 112–118
  survey sample, 109
  collaboration, 106, 107, 128–130, 131, 137–139, 143–145, 153, 182, 188
    group work, 50, 92–93, 94–96, 97, 101, 122, 138, 163, 164–165, 166, 181
    peer support, 105, 116–118, **117**, 120–121, 129, 165
    social networks, 55, 125, 153, 156, 163, 168–169 *see also* student organizations
    socio-emotional support, 121–123, 163, 165–166
  communication, 120, 143, 183–186
  community colleges, 11, 48–49
    advising, 168, 169, 170
    articulation agreement (Florida 2+2 program), 11
    class size, 157–158, 167
    collaboration, 164–166
    cost, 156
    faculty availability and support, 158–162, 168
    "feeder" schools, 11–12, 49, 147
    gender demographics, 149–150
    group work, 163, 164–165, 166
    instructors, 161–162
    mentorship, 163, 166, 169, 170, 173
    non-school obligations, 169–170
    participatory methods, 166–167
    pathway to four-year institution, 147, 155, 167
    qualitative methods, 147–148, 153–155
    racial demographics, 148
    reciprocity agreements, 49
    sample, 12, 147–148
    social networks, 153, 156–157, 163, 168–169
    socio-economic demographics, 150
    socio-emotional support, 163, 165–166
    student organizations, 162, 163–164
    student uncertainty, 170–171
    transfer students, 149, 151–153
  cooperation and harmony scales
    diversity, 110
    helpfulness, 110
  course sequencing, 120, 185, 188
  cultural capital *see* capital
  culture
    advising, 143, 145
    collaboration, 128–130, 131, 137–139, 143–144, 145
    collegiality, 130–131, 142
    conceptual framework, 3–4, 3
    definition of, 7
    departmental ranking, 130, 131, 134–135
    diversity, 130, 141–142
    doctoral program, 130, 131, 133–134, 145

engineering department culture, 127
faculty support, 125, 128, 131, 136–137
institutional support, 145–146
mentorship, 128–129, 134, 136, 145–146
personal strategies, 128
political economy, 127, 128, 145
practice theory, 127, 128
program efficacy, 129
real world application, 131, 139–141
research, 128, 130, 131, 132, 135
student support, 128, 143–144
student organizations, 143, 145, 146
teaching, 128, 130, 132–133, 135

Demaree, Robert G. *see also* r_{wg} *under* methods
dual enrollment, 87

economic disadvantage *see* political economy
enacted values, 130, 134
engineering departments
  gender equity of, 24
  racial diversity of, 22–24
  see under individual university names
engineering interest, 60–61, 62, 65–66, 67, 69–71, 81–86, 87, 88–89, 102–103, 174–176, 178, 180, 189–190
espoused values, 17, 130, 134

faculty research, 118, 125, 128–129, 130–137, 145–146, 159, 181–182, 183, 184
faculty support, 110, 112–118, **113**, **115**, **117**, 121, 125, 128, 131, 136–137, 157, 184
financial aid, 4
fit, **3**, 6–7, 16, 18, 53–57, 60, 65, 73–78, 79, 105–106, 111, 112, **113**, **115**, 116–117, 118, 122–125, 164, 188–189
  person-environment (PE), 3, 6–7, 55, 57
  social, 18, 54, 75–78, 79, 105–106, 111, 112, **113**, **115**, 116–117, 118, 122–125, 164, 188
Florida A&M University (FAMU) and Florida State University (FSU)
  campus ecology, 33–35
  history of, 27–29
  joint program with Florida State University, 29
  program efficacy, 31–32
  student demographics, 29–31
Florida International University (FIU)
  campus ecology, 38–40
  history of, 35–36
  program efficacy, 36–38
  student demographics, 36
Florida State University (FSU) *see* Florida A&M University
Florida State University System (SUS), 3

Giddens, Anthony, 3
graduation rates *see* program efficacy

Hall, Roberta, 107
Hewitt, Nancy, 54, 91, 92, 103, 107, 174
hybridity, 4

instrument development, 12
integration scales
  engagement, 111
  fit, 111, 112, 117, 118, 122–125
  importance, 111, 113
  integration, 110–111, 116
intent-to-leave, 105, 111, 112, 116, 117, 118

James, Lawrence *see also* r_{wg} *under* methods

Lewin, Kurt (1890–1947), 108

May, Gary, 129
mentorship
  faculty, 119, 120, 128–129, 134, 136, 145–146, 162, 173, 175, 183, 186–188, 190

mentorship—*Continued*
  peer, 188–189, 163, 166, 169–170, 188–189
  personal, 82, 83, 85, 175
methods
  ATLAS ti, 14
  climate survey, 12, 13, 15, 16, 18, 84, 86, 102, 108–118
  climate survey, inter-rater agreement, 108
  community college, 153–155
  constant comparative method, 112
  data collection, 13–14
  ethnographic, 26–27
  interviews, 13–14, 57–58, 81, 108–109, 128, 130, 153–154
  interview coding, 14
  participants *see* sample
  participatory workshops, 166–167
  recruitment, 13
  r_{wg} statistic, 108, 111, 112
  selection, 113
multi-level model, **3**

National Science Foundation, 2, 10

office hours, 118–119, 135, 136–137, 159, 168, 170

participants *see* sample
participatory methods, 154–155, 166–167
pedagogy, faculty perspectives
  lectures, 93–94, 97
  interactive teaching, 94–96, 97
  real world applications, 96–98
pedagogy, student perspectives
  lectures, 98–99, 103
  interactive learning, 99–100, 103–104
  real world applications, 100–101, 103–104
  student adaptations, 101
persisters, 53, 56, 58–60, 67, 68, 71, 73–76, 79, 189

person-environment (PE) fit *see* fit
preparation, academic, 5, 11, 41, 54, 55, 58–60, 62–66, 67–69, 73, 79, 86–91, 102–104, 124, 152–153, 157, 174, 176–179, 189
program efficacy, **3**, 8–9, 24–26, 106–108, 109, 111–112, 118, 122, 125, 128–130, 170, 178
  FAMU-FSU, 31–32
  FIU, 36–38
  UF, 42–43
  USF, 46–47

real world application, 92, 96–97, 100–101, 103–104, 138, 139–141, 160, 180, 181–183
recruitment *see* methods
research participants *see* sample
research team, 9
retention *see* program efficacy
role models, 84–86, 104, 110, **117**, 119, 175, 184, 188

sample
  community college interviews, 12, 147–148
  qualitative interviews, 13, 27
  participatory workshops, 166
  survey, 109
  switchers, 58
Sandler, Bernice, 107
  see also chilly climate
Seymour, Elaine, 54, 91, 92, 103, 107, 174
scales, *see under individual scale names*
science, technology, engineering, technology (STEM), 1–2
  employment, 1–2
  undergraduate STEM degrees, 2
  labor force demographics, 2
scientists women and minorities, scarcity of, 1
selection criteria *see* methods
social capital *see* capital

social fit *see* fit
stereotype threat, 54–55
student organizations, 106, 107, 125, 143, 145, 146, 162, 163–164, 188–189
   Florida-Georgia Louis Stokes Alliance for Minority Participation (FGLSAMP), 155, 166, 169
   National Society for Black Engineers (NSBE), 43, 125, 145, 154, 162, 165, 189
   Society for Women Engineers (SWE), 40, 145, 189,
study limitations, 15
support and facilitation scales, 109–110
   faculty support, 110, 112, 116, 117, 118
   involvement, 109–110, 112
   institutional support, 110, 112, 116, 118–119
survey *see under individual scale names*; *see* climate; methods
switchers
   academic preparation, 54, 55, 58–60, 62–66, 67–69, 73, 79
   capital, 54, 55, 56, 64, 73, 75, 79
   definition of, 53
   engineering interest, 60–61, 62, 65–66, 67, 69–71
   fit, 53–57, 60, 65, 73–78, 79
   political economy, 55–56

   practice theory, 56–57
   preparation, academic, 54, 55, 58–60, 62–66, 67–69, 73, 79
   sample, 58
   stereotype threat, 54–55
   student organizations, 40–43
   symbolic capital *see capital*

theoretical frameworks
   fit, 3–4, **3**, 6–7, 16, 18, 53–57, 60, 65, 73–78, 79, 105–106, 111, 112, **113**, **115**, 116–117, 118, 122–125, 164, 188–189
   political economy, 3–5, 55–56, 127, 128, 145, 152
   practice theory, 3–4, 5–6, 56–57, 107, 127, 128, 152
Tinto, Vincent, 106, 123
   integration process, 107

University of Florida (UF)
   campus ecology, 43–44
   history of, 40–41
   program efficacy, 36–38
   student demographics, 41–42
University of South Florida (USF)
   campus ecology, 47–48
   history of, 44–45
   program efficacy, 46–47
   student demographics, 45–46

Willis, Paul, 5
Wolf, Gerrit *see also* r_{wg} *under* methods

Also by Kathryn M. Borman:

Sociological Perspectives on No Child Left Behind, co-edited with Sadovnik, O'Day, and Bohrnstedt, 2007.

Education Politics and Policy in Florida, 2006.

The Encyclopedia of the High School, co-edited with Cahill and Cotner, 2006.

Examining Comprehensive School Reform, co-edited with Aladjem, 2006.

Meaningful Urban Education Reform: Confronting the Learning Crisis in Mathematics and Science, 2005.

Ethnic Diversity in Communities and Schools: Recognizing and Building on Strengths, 1998.

Youth Experience and Development: Social Influences and Educational Challenges, co-edited with Schneider, 1998.

Women and Work: A Reader, co-authored with Dubeck, 1997.

The Handbook of Women and Work, co-edited with Dubeck, 1996.

Implementing Educational Reform: Sociological Perspectives on Educational Policy, co-edited with Cookson, Sadovnik, and Spade, 1996.

Restructuring Education: Issues and Strategies for Communities Schools and Universities, co-edited with Yinger, 1996.

Changing Schools; Recapturing the Past or Inventing the Future?, co-edited with Greenman, 1994.

Investing in the U.S. Schools: Directions for Educational Policy, co-edited with Jones, 1994.

Effective Schooling of Economically Disadvantaged Students, co-edited with Johnston, 1992.

The First Real Job: A Study of Young Workers, 1991

Contemporary Issues in American Education, co-edited with Swami and Wagstaff, 1990.

Work Experience and Psychological Development through the Life Cycle, co-edited with Mortimer, 1988.

Becoming a Worker, co-edited with Reisman, 1986.

Schools in Central Cities, co-edited with Spring, 1984.

Women in the Workplace: Effects on Families, co-edited with Quarm and Gideonse, 1984.

The Social Life of Children in a Changing Society, 1982.

Also by *Rhoda H. Halperin*:

Who's School Is It? Women, Children, Memory and Practice in the City, 2006.

The Teacup Ministry and Other Stories: Subtle Boundaries of Class, 2001. *Economies across Cultures: Towards a Comparative Science of the Economy*, 1998.

Practicing Community: Class Culture and Power in an Urban Neighborhood, 1998.

Cultural Economies Past and Present, 1994.

The Livelihood of Kin: Making Ends Meet the Kentucky Way, 1991.

GPSR Compliance

The European Union's (EU) General Product Safety Regulation (GPSR) is a set of rules that requires consumer products to be safe and our obligations to ensure this.

If you have any concerns about our products, you can contact us on

ProductSafety@springernature.com

In case Publisher is established outside the EU, the EU authorized representative is:

Springer Nature Customer Service Center GmbH
Europaplatz 3
69115 Heidelberg, Germany

www.ingramcontent.com/pod-product-compliance
Lightning Source LLC
LaVergne TN
LVHW011818060526
838200LV00053B/3824